THE FOSSIL WOMAN
A LIFE OF MARY ANNING

Tom Sharpe is a geologist who has worked as a national museum curator, university lecturer, and expedition travel guide, mainly in the polar regions. He is a Fellow of the Geological Society and has been Chair of its History of Geology Group and Geological Curators' Group, as well as a trustee of Lyme Regis Museum – part of whose building stands on the site of the house in which Mary Anning spent her childhood. He has published on the geologists William Smith and Henry De la Beche and on the history of the geological exploration of Antarctica. He first encountered Mary Anning in *The Observer's Book of Geology* when he was twelve.

FOLLOWING PAGE
Sketch by Mary Anning of the complete *Plesiosaurus* which she discovered in December 1823, in a letter of 26 December 1823 to fossil collector Sir Henry Bunbury.

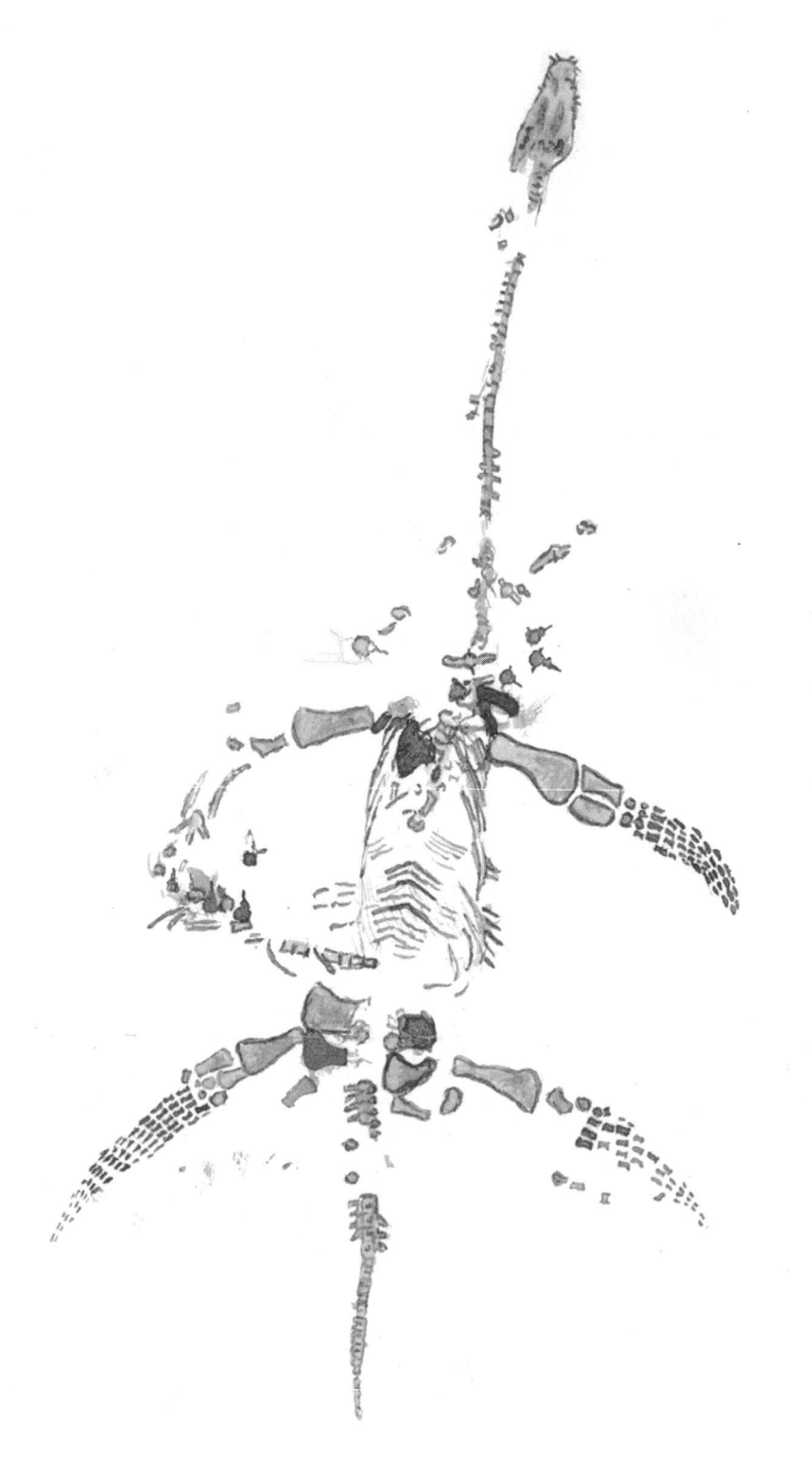

THE FOSSIL WOMAN
A Life of Mary Anning

TOM SHARPE

THE DOVECOTE PRESS

To
C, G & G

First published in 2020, and as a paperback in 2021
Reprinted 2022
by The Dovecote Press Ltd
Holdscroft Farm, Marshwood, Bridport, Dorset DT6 5QL

ISBN 978-1-8384735-0-1
Text © Tom Sharpe

Tom Sharpe has asserted his rights under the
Copyright, Designs and Patent Act 1988 to be identified as author of this work

Typeset in Sabon and designed by The Dovecote Press Ltd
Printed and bound in Cornwall by TJ Books, Padstow
All papers used by The Dovecote Press are natural, recyclable products
made from wood grown in sustainable, well-managed forests.

A cip catalogue record for this book is available from the British Library

All rights reserved

3 5 7 9 8 6 4 2

Contents

Introduction 7

Author's Note 10

1. 'Old Wonders and New Improvements' 12
2. The Girl Who Lived 23
3. The First Ichthyosaurs 33
4. Friends and Neighbours 42
5. The First Plesiosaurs 54
6. Decade of Discoveries 79
7. Mary and the Sea Dragons 99
8. Fossil Fish and Financial Calamity 124
9. 'I am well known throughout the whole of Europe' 136
10. Legacy and Legend 148

Mary Anning's Fossils 164

Who's Who 173

Notes, Sources & Further Reading 177

Acknowledgements 234

The Illustrations 235

Index 236

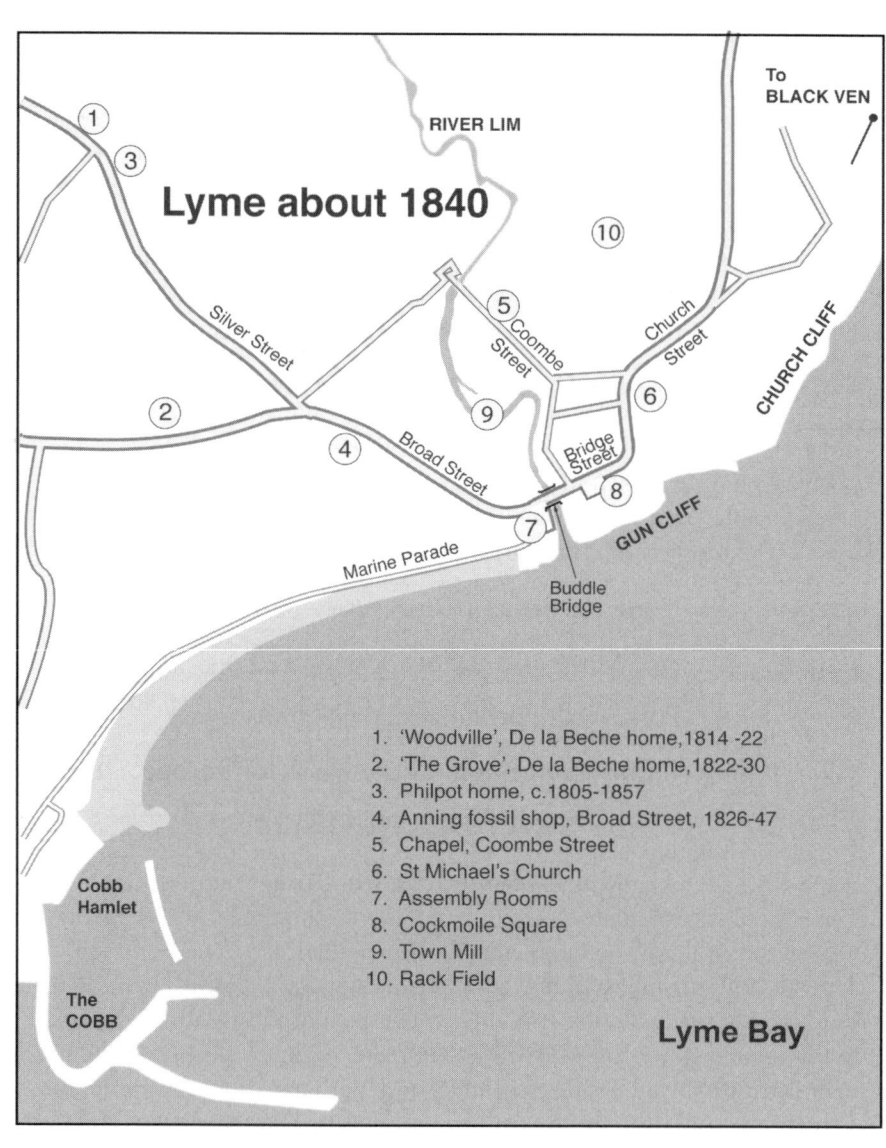

Introduction

IN FEBRUARY 1830 MARY ANNING wrote to the geologist William Buckland to say she was sending him a box of samples, amongst them some fossil reptile faeces, known as coprolites, and describing a new fossil skeleton she had recently uncovered in the cliffs near Lyme Regis in Dorset. The letter was sold at auction by Sotheby's in August 2020 for over £100,000 – ten times its original estimate.

Until recently it could safely be assumed that Mary Anning was the most famous person many of us had never heard of. Thirty years old when she wrote the letter, a working class woman who never married, Anning made her living as a 'fossilist', a finder and seller of fossils, an unusual occupation, especially for a woman. Along with her brother and mother, Mary Anning's remarkable discoveries drew scientific attention to a previously unknown world of extinct marine reptiles preserved in the cliffs and foreshore around Lyme. These strange animals challenged what we then understood about the history of life on earth.

Mary Anning's skill as a fossil discoverer was also matched by her expertise in preparing her specimens – extracting the fossils from their embedding rock matrix – and by her knowledge and understanding of their anatomy. When still in her early twenties she had become a significant figure in the geological community of the early nineteenth century. Despite a limited education, and as the letter to William Buckland makes clear, she confidently corresponded with some of the leading scientists in Europe. Anyone with an interest in fossils went to Lyme Regis to meet her, to negotiate the purchase of specimens, to accompany her on excursions out onto the rocks along the shore and to learn from her. Many of the fossils she found are now held in museum collections around the world, and are still used for scientific research two hundred years after Anning first uncovered them.

Although famed and celebrated during her lifetime, and written about long after her death, Mary Anning's name gradually slipped into relative obscurity, known only to geologists, fossil collectors and historians of science.

But the tide is turning, and it is now widely accepted that her considerable achievements merit the public recognition given to other great women of the nineteenth century, such as Jane Austen or Florence Nightingale. A few years ago the Royal Society included her on a list of the ten most influential women in the history of British science, and in 2018 the Bank of England considered her portrait for the new £50 note.

What makes Mary Anning's renaissance even more remarkable is that there is much we still don't know about her. The exact places of both her birth and burial remain shrouded in uncertainty. There is no single archive for biographers to draw on. Her few surviving papers were dispersed in the 1880s and many were subsequently lost or remain unrecognised in the archives of her contemporaries. We have some correspondence and notes made by Anning herself; some references to her by her friends, acquaintances and visitors to Lyme in diaries and journals; and a few official records. Scant pickings for a biographer to weave into a narrative.

Much of what we do know about Mary Anning is due to the research of William Dickson Lang, Keeper of Geology at the British Museum (Natural History) from 1928 until 1938, and a not inconsiderable scientist in his own right. Lang had conducted a detailed study of the rocks exposed on the coast near Lyme, and on his retirement in 1938 moved to Charmouth, a neighbouring village. Between 1935 and 1963, as he unearthed previously unknown letters and diaries, Lang published a series of papers about Anning, mostly in the *Proceedings of the Dorset Natural History & Archaeological Society*, and these are still key source material for anyone writing about Anning today.

We also have the writings of Lyme schoolmaster and historian George Roberts, a contemporary of Mary Anning, whose school was situated on Broad Street opposite the home and shop of Mary and her mother. He knew the Annings well and wrote about Mary in several of his books about the history of Lyme and Charmouth. He was the probable source of many reports about the Annings' fossil discoveries published in provincial newspapers, effectively acting as Mary Anning's publicity agent.[1]

While we can speculate based on our reading and interpretation of those few sources that do exist, we must tread with caution, especially as we are looking back two hundred years through twenty-first century eyes, thus running the risk of grafting our own biases and agendas onto the bare bones of Mary Anning's story. The life of a poor, working class woman in rural Dorset within the strictly stratified and patriarchal society of the early nineteenth century is far removed from the experience of most today.

The last ten years or so have seen much published that is new, mainly in academic journals, about Anning and her times, leading to the questioning and reassessing of some of the oft-quoted sources about her. The telling and retelling of Mary Anning's story over the years has meant some aspects of her life becoming accepted as fact when there is scant evidence to support them. Much of the more recent research has been carried out by two distinguished British historians of geology, Professor Hugh S. Torrens and Dr Michael A. Taylor, and I am much in their debt.

There is little doubt that Mary Anning's star is in the ascendant – and not before time. In writing this book I have inevitably spent a considerable time thinking about her character and all that she achieved. By any standards she was a remarkable woman, and my admiration for her has grown the more I have learnt. Feminist icon, gifted scientist, the miraculous survivor of a lightning strike, a spinster doomed never to find someone to share her life with, or just the name by which she was known in her lifetime, 'the fossil woman', she is not an easy person to pin down. Such is the air of mystery that still surrounds her she has recently taken centre stage in several novels, plays and films which skilfully blur fact and fiction. The Sotheby's sale of her letter makes me even more certain that a fresh look at her life is overdue, to separate myth from reality, to assess her contribution to science, and to examine her relationship – as a woman – with her fellow pioneers – all men – of the then fledgling science of geology at a time when her discoveries were turning our knowledge of life on earth on its head.

Author's Note

UNITS OF CURRENCY AND MEASUREMENT

In Anning's day, indeed up until the adoption of decimal currency in 1971, the UK monetary system used pounds, shillings and pence. Twenty shillings equalled one pound, and twelve pence equalled one shilling, with 240 pence to the pound. Amounts were expressed as, for example, £1. 12. 6 or £1 12s 6d for one pound, twelve shillings and sixpence ('d' was the abbreviation for penny). A penny was subdivided into two halfpennies or four farthings. Also in common use in the early nineteenth century were guineas; one guinea was one pound and one shilling, so five guineas was £5. 5s. Coins in circulation included sovereigns, which were gold £1 coins; crowns, which were five shillings; and halfcrowns. 'Half a crown' which Mary Anning was given for her first fossil, was 2s 6d.

It is difficult, perhaps even meaningless, to compare sums of money from the early nineteenth century with what might be their early twenty-first century equivalents, but some form of comparison is necessary, otherwise we cannot appreciate what the sale of fossils meant for Mary Anning's income. A currency converter on the website of the National Archives gives the very roughest idea of monetary values. For example, the halfcrown given to Mary Anning would be about five or six pounds today, which seems about right. The £23 the Annings received for their first ichthyosaur would be about £1,000 today; perhaps a more useful comparison is that £23 in 1810 would buy two horses or four cows and was the equivalent of 150 days wages for a skilled tradesman. One hundred pounds in 1820 was over 650 days wages for a skilled worker and about £5,700 today. These are broadly the sorts of sums for which Anning specimens were being sold at the time. However, there are so many socio-economic factors that need to be considered: the desperate poverty of rural Dorset in the early nineteenth century, the deeply hierarchical society and the extreme imbalance in the distribution of wealth. It has been suggested that multiplying the figures by at least 200 might give us a better idea of what the Anning prices might be today. So the £23 ichthyosaur might be £5,000, which seems a little more reasonable, although still perhaps a considerable underestimate.[1.]

Measurements of distance and length used in the nineteenth century, and still much in use today in parallel with the metric system, are miles, yards, feet and inches. A mile is the equivalent of 1.6 kilometres; a yard is about 91 centimetres, a foot is 30 cm and an inch is 2.5 cm. You'll find a mix of imperial and metric measurements used throughout this book.

FOSSIL NAMES

The names of fossils, like those of modern plants and animals, comprise two parts, a genus name and a species name which, by convention today, are italicised. The genus name begins with an upper case letter, while the species name is all lower case, for example, *Ichthyosaurus communis* or *Plesiosaurus dolichodeirus*. Where original sources, which are often handwritten, are quoted the names are not italicised, but they may be so in quotes from published sources. The names are derived from Greek and Latin roots and are created by the scientists who are describing a new genus or species for the first time.

QUOTATIONS

I have made extensive use of quotations from original sources – mainly letters, journals and both scientific and popular publications of the period. I think this helps to develop a feel for the vocabulary and terminology of the time, and the way in which Mary Anning and her contemporaries expressed themselves. In these quotations, I have tried to retain the original spelling and punctuation. In the case of letters written by Anning, at times her phonetic spellings hint at her West Dorset accent.

AMMONITE CHAPTER HEADING

The ammonite illustrated at the head of each chapter is a common fossil from the Charmouth Mudstone Formation of the Lias of Lyme Regis, *Asteroceras obtusum*, first described by James Sowerby in *The Mineral Conchology* in 1817, and taken from William Buckland's 1836 *Bridgewater Treatise*.

TOM SHARPE

ONE

'Old Wonders and New Improvements'

In April 2014 a metal detectorist called Phil Godwin was searching the beach below Church Cliffs in Lyme Regis when his device suddenly alerted him to a possible find. Carefully scraping away the pebbles and sand, he found a small brass disc with the crudely stamped inscription MARY ANNING/ MDCCCX on one side and LYME/REGIS/AGE XI on the other. It was only an inch in diameter, a random but remarkable discovery – the equivalent of the needle in a haystack. Phil found it following a winter of storms, wind and wave washing away the sand to expose the underlying rock ledges at the foot of the cliffs. No similar disc had ever been discovered. In all probability it was specially made for Mary Anning by her father to mark her eleventh birthday on 21 May 1810, and then lost, perhaps many years later. But nowhere could have been more fitting: these unstable cliffs had yielded her most celebrated finds.

Mary Anning was born, lived, and died in Lyme Regis. This was her world, and as a child she would have grown used to the great south-westerly gales that swept across the broad curve of Lyme Bay, from Start Point in the west to the distant outline of the Isle of Portland in the east. She might also have learnt something of Lyme's changing fortunes, of its glory days as a medieval port trading in wine and wool, of the Royal Charter granted it by Edward I adding 'Regis' to the name by which it is still known locally. With no natural harbour, the ships that dropped anchor in Lyme were dependent for their protection on what is now its most famous landmark, the sinuous curve of a stone rampart known as the Cobb. The sea remained the town's principal link to the rest of the country until the mid eighteenth century construction of a new turnpike road which climbed steeply uphill to join the main stagecoach route between London and Exeter. Prior to that, the only access from inland was on foot or by pack-horse down steep, rutted lanes.

There are two distinct parts to Lyme: Cobb hamlet, centred around the harbour, and the town, about half a mile to the northeast, clustered along the steep valley sides of the little River Lim from which the cloth mills took their power and water. Separating the town and its harbour is an area of unstable and landslipped ground unsuitable for building, and which is now the parks of Lister and Langmoor Gardens. Along the shore at its foot, the promenade of Marine Parade links the Cobb with the town. Initially known as The Walk, this was begun in 1771 by the political philosopher Thomas Hollis as being more suitable than the beach for Lyme's increasingly fashionable visitors, to which horse-drawn carts remained relegated for their journeys between town and Cobb. Lyme's debt to Thomas Hollis is considerable. Staunchly Radical (he named one of the fields on his Corscombe estate 'Stuart' so he could have the pleasure of annually beheading it at harvest; he was also a great benefactor to Harvard University), he either bought up and demolished Lyme's poorer tenements, replacing them with new houses, or persuaded their owners to modernise.[1.]

In the 18th century, the health benefits of bathing in the sea and even of drinking sea water began to be promoted by doctors, notably the Sussex physician Richard Russell who popularised the treatment in Brighton. The gentry and nobility, and a developing middle class – those with money and time to spare – descended on coastal towns such as Brighton, Weymouth and Lyme Regis, especially after the construction of the new road improved access.[2.]

These new visitors were Lyme's lifeline. The town was in decline, not helped by it being a Rotten Borough whose two Parliamentary seats had been in the pocket of a Bristol family called Fane since the 1730s. In John Fowles's *A Short History of Lyme Regis* he describes them as running the town with 'a ruthless contempt for its ancient privileges and freedoms'. 'I have bought you all, and by God I will sell you all,' said the head of the family, the Earl of Westmorland, when addressing the Corporation in 1766. His son Henry was Lyme's MP for 30 years until 1802, yet never said a word in Parliament. Successive Fanes corrupted the Customs service, failed to keep the sea walls in repair, and effectively 'fixed' elections by restricting the electorate to their cronies. One Fane or another represented Lyme for much of Mary Anning's life, only losing control when the Reform Act of 1832 reduced it to a single member of parliament.[3.]

If political chicanery was one brake on Lyme's revival, the other was the Cobb. Merchant ships had grown too large for it, the wine trade with France had been interrupted by war, and the focus of the woollen industry had shifted to the northern mill towns. Although its glory days were over, there was still a shipbuilding yard, owned when Mary Anning was a child by Henry Chard. The 40 vessels he built were a mixture of sloops, cutters and luggers, averaging about 50 tons. They were much used by local smugglers, and a handful were sunk or captured by the French. Cobb hamlet was largely made up of warehouses, timber yards, fishing nets and lobster pots – adding a nautical flavour for visitors walking Marine Parade.[4]

The inclination to take a daily dip in the bracing waters of the English Channel was just what Lyme needed to restore its fortunes. As Lyme historian Cyril Wanklyn wrote in *Lyme Regis: a retrospect* (1927), mid eighteenth century Lyme was casting off 'its homespun, but rather threadbare, industrial clothes ... [to] rise up again in the more attractive, if less serviceable, apparel of a watering-place'.[5]

While Brighton and Weymouth, with their royal connections to the Prince Regent and King George III, attracted more aristocratic visitors, Lyme appealed to the new middle and professional classes. 'Lodgings and boarding at Lyme are not merely reasonable, they are even cheap ... it is frequented by persons in the middle class of life, who go there, not always in search of their lost health, but ... to heal their wounded fortunes, or to replenish their exhausted revenues', wrote John Feltham in 1806 in his *Guide to all the watering and sea-bathing places*.[6]

Harriette Wilson, mistress of the 1st Duke of Wellington (amongst others) in her *Memoirs* recalled a visit to Lyme that same year

> 'Lyme Regis is a sort of Brighton in miniature, all bustle and confusion, assembly-rooms, donkey-riding, raffling &c. &c. It was sixpence per night to attend the assemblies, and much cheaper if paid by the season ... They were very pleasantly situated near the sea ... but the society was chiefly composed of people of very small independent fortunes, who for economy had settled at Lyme Regis; or such as required sea-bathing.'[7]

Lyme benefited from its proximity to the more celebrated spa at Bath, where visitors had first taken the waters in Roman times, and whose popularity grew rapidly in the late eighteenth and early nineteenth centuries. Bath and its Assembly Rooms became a centre for fashionable society, whose members would exchange the city's elegant Georgian terraces for Lyme's lodging houses after the Bath 'season', which ran from October to early June. Bath lay 60 miles to the north, or an uncomfortable eight hour coach journey, and in his 1823 *History of Lyme-Regis*, George Roberts describes this migration:

> 'As soon as the season at Bath is over, of which Lyme has been humorously considered as a safety-valve, the company who have houses in Lyme return from their brumal [winter] migration, when gaiety soon commences, and continues without intermission till the end of autumn.'[8]

Emulating Bath, Lyme had its own Assembly Rooms, constructed by

Thomas Hollis and completed in 1775. Although small, these became the social hub for both visitors and residents alike. Describing the Rooms, George Roberts attributed 'the revival and popularity Lyme has acquired' to their construction and was probably correct to suggest that 'they may have rescued the town from impending ruin.' During the season, balls were held every Tuesday, of which some were 'not unfrequently attended by two hundred fashionables ... The superiority over similar establishments, and social intercourse at the Rooms, are universally allowed by visitors'. Feltham, in his guidebook, declared that 'the rooms at Lyme frequently exhibit as cheerful countenances as are to be seen at Bath or Brighton.'

One of these cheerful countenances, which might also have been seen in the Assembly Rooms in Bath, was that of the yet-to-be published novelist Jane Austen, who came to stay in Lyme in 1803 and 1804 when in her late twenties. In a letter written to her sister Cassandra during her second visit in September 1804, she recounts spending her evenings at the Assembly Rooms playing cards and warding off the unwelcome attention of the 'odd-looking' son of an Irish peer – 'bold, queer-looking people, just fit to be quality at Lyme'. The town certainly left an impression; in *Persuasion*, begun over a decade later in 1815, she gives a description worthy of any guidebook:

> 'as there is nothing to admire in the buildings themselves, the remarkable situation of the town, the principal street almost hurrying into the water, the walk to the Cobb, skirting round the pleasant little bay, which, in the season, is animated with bathing machines and company; the Cobb itself, its old wonders and new improvements, with the very beautiful line of cliffs stretching out to the east of the town, are what the stranger's eye will seek; and a very strange stranger it must be, who does not see charms in the immediate environs of Lyme, to make him wish to know it better.'[9.]

A less favourable impression was left by a local cabinet-maker called by the Austens to their rented house on Broad Street to estimate the cost of repairing a piece of furniture.

> 'I have written to Mr Pyne [the owner of the house] on the subject of the broken lid; it was valued by Anning here we were told at five shillings, & as that appeared to us beyond the value of all the furniture in the room together, we have referred ourselves to the owner.'[10.]

This unacceptable estimate was provided by Richard Anning, carpenter of Lyme, and fossil-selling father to a then-five year old Mary. Although the fossils were a sideline to his cabinet-making, it was Richard who initially fostered Lyme's reputation as a prime destination for those interested in geology and palaeontology. By the mid 1830s, George Roberts could proudly state that 'The *fossilist* and *geologist* look upon this place as the sportsman does upon Melton Mowbray'.[11]

The Leicestershire market town's fame as the cradle of foxhunting made it much bigger than Lyme. But as the number of visitors rose, so too did Lyme's population, from 1451 in 1801 to 2756 forty years later. Even today, especially during the winter, it remains a small coastal town of under 4,000.

LYME ROCKS

Development in Lyme has long been hampered by its position astride the narrow, steep-sided valley cut by the River Lim. The rocks along this section of coast render the slopes unstable and liable to slip, so the town has been squeezed into a geological corset. To the west lie the large landslips of Ware Cliff and the Undercliff, while to the east the landslips of The Spittles and Black Ven are amongst the largest in Europe.

The instability of the ground around Lyme is due to the interlayering of permeable limestones and impermeable mudstones and to the tilt, what geologists call the dip, of the beds of rocks. During heavy rain, water soaks through the limestones and, unable to pass though the mudstones, seeps along the border between the two rock types, causing the overlying rocks to collapse.

On the coast, the softer mudstones are also more readily attacked by the sea, especially when storms batter and undermine the base of the cliffs, causing rock falls and erosion. Where the cliffs are unprotected by sea defences, it has been estimated that every year about 18 inches (50 cm) are lost, with the clifftop retreating inland. This crumbling away of the coast has been a problem at Lyme for centuries, exacerbated in the nineteenth century by quarrying of limestones on the foreshore for the manufacture of cement and stucco.

The rocks around Lyme Regis are a sequence of layered sedimentary rocks deposited under the sea about two hundred million years ago during a period of time now called the Jurassic. In the early nineteenth century, however, the name by which they were commonly known was 'Lias' or 'Blue Lias', hinting at their grey-blue colour. The origin of the term is unclear, but most likely originates in the pronunciation of the word 'layers' in the local West Country dialect. 'Lias' is still used by British geologists today to refer to rocks of the early Jurassic Period, not just in Dorset, but also where similar rocks of the same age crop out, including Somerset and Glamorgan, through the Midlands to Lincolnshire and Yorkshire and north in the Hebrides.[12]

Immediately to the east of Lyme the Lias is divided into two, a lower sequence of limestones and mudstones called the Blue Lias Formation, about 85 feet (26 m) thick, and an upper sequence of mainly mudstone rocks, the Charmouth Mudstone Formation, about 400 feet (120 m) thick. The Lias rocks are tilted gently towards the east. On top of these, and forming the upper part of the cliffs, are layers of 100 million-year-old flat-lying younger rocks from the Cretaceous Period. The Cretaceous rocks are composed of about 40 feet (12 m) of clays called the Gault Formation, on which rest a sequence of sandstones called the Upper Greensand Formation.

The Blue Lias Formation forms the cliffs to the west of the Cobb but farther west into East Devon older red sandstones of the earlier Triassic Period appear beneath it. The geology of the East Devon and Dorset coast is spectacularly exposed from these older red rocks in East Devon with successively younger rocks appearing eastwards all the way to Studland Bay in Dorset. In recognition of this beautifully exposed span of 185 million years of geological time and of its importance to the history of geology this forty mile stretch of coastline was designated the Jurassic Coast World Heritage Site by UNESCO in 2001.[13]

LIFE IN THE ROCKS

Two hundred million years ago, Britain lay at a lower latitude than its present position, with southern England at about 35° north, roughly the latitude of the Mediterranean today. A tepid, shallow sea covered much of the country apart from the upland areas of the west and north. The Lias sea teemed with life; bivalve molluscs lived on and in the soft muds of the sea floor on which sea-snails – gastropods – crawled and grazed. In the waters above them swam other mollusc groups, squid-like belemnites and coiled-shelled ammonites as well as many different forms of fish. Wood from nearby land occasionally drifted by, sometimes with the long stems of crinoids, plant-like echinoderms, attached, before becoming waterlogged and sinking to the sea bed. The monarchs of this prehistoric world were the marine reptiles, the round-bodied, long-necked plesiosaurs and the dolphin-like ichthyosaurs, some of which were huge. With skulls up to six feet or so long, and wide jaws armed with hundreds of pointed teeth, ichthyosaurs were the top predators of the Lias sea. Overhead, bat-like reptiles, the pterosaurs, flew through the warm air.

That we know about these animals is due to their preservation as fossils in the rocks. Fossils form when an animal – or plant – dies and becomes buried in sediment like sand, silt or mud. Soft tissues usually rot and decay, leaving behind only hard parts such as shells, bones or teeth. Animals, like marine worms, for example, which have no shell, bone or other hard parts are extremely rare in the fossil record. Over time, as layers of sediment build up, the remains of the organism are buried deeper and deeper. As water in the sediment is squeezed out by the weight of the overlying layers, grains of sand or silt become cemented together, and eventually, what was soft sediment becomes hard sedimentary rock. During this process, the original material of shell or bone may be replaced by other minerals, or even dissolved away completely to leave only an impression in the rock.

Fossils represent only a tiny fraction of the organisms that live in a particular environment, and a sample biased towards those with hard parts. Most die and decay or are eaten or scavenged before they have a chance to be buried and preserved. Animal skeletons can be broken up and the bones dispersed before burial, so the preservation of a complete and intact skeleton is a rare event. Even once fossils were formed, some did not survive. Before the oceans and land masses settled into their present positions the movement of tectonic plates caused continents to collide. Some rocks caught up in these collisions were subjected to such great heat and pressure that they recrystallised as metamorphic rocks, and the fossils disappeared. Rocks once sea floor mud became land.

Rivers bring sand, silt and mud eroded from the land into the sea, or the erosion of coastal rocks can provide the material for new sediment. Sand becomes sandstone, silt siltstone and mud – clay – mudstone. In some situations the mud can be rich in lime, calcium carbonate, and form a limy mudstone called marl, whilst in warm, tropical or subtropical waters lime can be precipitated to form limestone.

Organisms which live in the sea stand the best chance of becoming fossils as the sea is where sediments naturally accumulate. But their survival is subject to all sorts of hazards, and only a tiny fraction become sufficiently buried to embark on the long journey into fossilisation. Even if an organism survives the processes of burial, fossilisation, and uplift, hundreds of millions of years later its fossil might be destroyed by erosion as the sea eats away at the coastal cliffs. Or even lost in an instant by the misplaced strike of a fossil collector's hammer.

Land animals, on the other hand, are unlikely to be preserved as fossils as the terrestrial environment is not one where sediments are laid down. Rather, it is where rocks are eroded and become the source of the sediments forming in the sea. There are exceptions, of course, and fossils can form in river, lake and swamp deposits. It is also possible to find fossils of land animals and plants in marine sediments if their remains are washed out to sea and buried there. They are rare, but even insect and dinosaur fossils have been found in the marine Lias rocks at Charmouth. Fossil wood – driftwood – is not uncommon, often found with fossil marine organisms still attached.[14]

Several hundred years of collecting and research along the coast at Lyme have provided us with a good picture of life in the Lias seas. The shoreline here continues to be a source of new specimens. Venture onto the beach at Lyme, especially the foreshore to the east of the town and you will hear the tap-tap of hammer on rock. On a summer's day, family groups may be huddled around beach boulders trying to extract a fossil fragment; larger groups on an organised fossil walk, well-served by the sharp and experienced eyes of a local guide. There may be university students examining the rock sequence as part of their course; and there will always be a few solitary souls searching assiduously amongst the rocks. In summer, the serious collectors will be few; there is little to be found when the weather is quiet and the beach has been picked over by holidaymakers. In winter it is a different story. Wild, wet and windy weather deters all but the most dedicated collectors. For them these are the ideal conditions. After a storm has lashed the cliffs, there is a better chance of fossils being washed out or partly exposed, or of new cliff-falls and landslides revealing fresh rock. As a visitor to Lyme in 1844 noted:

> 'It is a piece of great good fortune for the collectors, when the heavy winter rains loosen and bring down large masses of the projecting coast. When such a fall takes place, the most splendid and rarest fossils are brought to light, and made accessible almost without labour on their part.'[15]

These were the conditions favoured by Mary Anning, as her customers knew. In January 1834, Oxford geologist William Buckland enquired of a friend in Lyme 'if the late violent storms have turned up any novelties upon the shores near Lyme or any good Specimens of the old School', while his wife expressed her hope that 'the late storms have been favourable to Mary Anning's Researches & that she has found something worth selling.'[16]

Reminiscences from later in the century recall,

> 'After every gale or heavy sea disturbance, the masculine figure (but always smiling face) of Mary Anning was seen with hammer and rush vasculum followed by her little black and white terrier dog, making towards the lias beds in search of the cast up or bared antedeluvian treasures with which her name has become so identified.'[17]

For the local collectors in Anning's day, and in ours, this hazardous environment was and is their workplace. There are few places in Britain where a professional fossil collector can hope to earn a living, and Lyme with its fossil-rich rocks is by far the best. Thanks to these fossil-hunters, fine quality specimens and even new species help add to our scientific knowledge and museum collections. Professional fossil collecting has a long pedigree in Lyme and Charmouth, going back to at least the late eighteenth century, and coinciding with the arrival of the first tourists to the area. With the high rate of coastal erosion and landslipping, and the extensive foreshore quarrying through the nineteenth century, there was a constant supply of additional material to meet the growing demand for souvenirs from tourists and visitors interested in the still youthful study of geology.

THE NEW GEOLOGY

Geology and the study of fossils, palaeontology, initially evolved as true sciences in the late eighteenth century – a mere 250 years ago. Although some philosophical debate as to the origin of rock strata and their fossils dates back to the ancient Greeks, it was only in the late seventeenth and early eighteenth centuries that fossils increasingly came to be accepted as the remains of once-living organisms. Some, like clam and oyster shells, were clearly recognisable as seashells, while those with no known living equivalents – ammonites and belemnites, for example – remained a mystery. It was obvious, too, that fossil bones were part of an animal; but what? Some were so huge it was conceivable they belonged to giants.

Rock strata containing fossil shells were recognised as marine deposits and some natural philosophers or *savants* – the terms 'scientist' and 'geologist' were yet to come into use – considered all rocks to have been formed under the sea, even those we now class as igneous rocks, formed from once molten material, like basalt lavas and granites. Others recognised that volcanoes played a part in the formation of certain rocks. Speculation and debate between the two camps raged through the closing and opening decades of the eighteenth and nineteenth centuries.

The Bible provided a convenient and widely accepted explanation for the shells and their embedding rock: they were the product of the Deluge, Noah's 'Flood'. Gradually belief in the 'Great Flood' became restricted to what were clearly more recent deposits of boulders and gravels. The fossils in the rocks, it was assumed, must have been the inhabitants of an antediluvian, pre-'Flood', world. The opening chapter of the Book of Genesis describes the six days God took to form heaven and earth and all living things, 'and on the seventh day God ended his work which he had made'. Theologians and scholars calculated that this had taken place about six thousand years earlier.

During the eighteenth century, the economic benefits from mining for minerals and their ores led to an increased interest in their origins. It began to be recognised that six thousand years was too short a period to build up the enormously thick layer of sedimentary rocks with which much of the planet is composed. In the 1780s, the Scottish doctor and farmer James Hutton proved that some rocks were formed by volcanic action, others by sedimentation. Widely regarded as the 'father' of modern geology, Hutton argued that the processes that take place today are identical to those that took place in the past, and occur over a long period of time. The six thousand years became tens of thousands, perhaps hundreds of thousands. Then stretched still further as some scientists began to think in terms of millions or hundreds of millions of years.

Until as recently as the 1950s it was impossible to date with any accuracy the age of a rock. Today, geologists measure tiny amounts of radioactive elements contained in certain kinds of rocks, in the same way that archaeology uses carbon14 to date organic material. We now know that the earth is about 4.6 billion years old and that the Jurassic Period began 201.3 million years ago.[18.]

While the actual age of a particular rock was unknown to geologists in the first half of the nineteenth century, they could decipher which were formed first, and which came later. In much of southern and eastern Britain this was reasonably straightforward as the pile of sedimentary rocks tilts gently towards the southeast. The rocks at the bottom lay in the northwest and could be found in Snowdonia in North Wales, and those at the top occurred around London and East Anglia.

In the 1790s a Bath-based surveyor called William Smith noticed that the rock strata were arranged in a regular sequence and that each stratum contained distinctive fossils, different from those in the rocks above or below. This allowed him to draw up a table of the order of strata around Bath and a geological map of the area. Using fossils to distinguish between different strata and identify where in the sequence each lay, Smith extended his mapping to cover the whole of southern Britain. His aim was to identify where coal, iron, limestone and other valuable materials required to power the Industrial Revolution could be found and, as an aid to farmers, to identify the soils which developed on different rocks. Smith's map was literally groundbreaking. The strata names used by Smith and other later geologists evolved through the nineteenth century into the geological column of time periods used internationally today.[19.]

LYME'S GEOLOGICAL QUARTET

In addition to a ready supply of fossils and a growing tourism industry, Lyme had a fortuitously close association with three men who were to be in the vanguard of the new interpretations of fossils and earth history: the eccentric

Oxford professor, the Reverend Dr William Buckland, born in nearby Axminster in 1784 and a frequent visitor to Lyme, both as child and adult; the reserved but brilliant clergyman, William Daniel Conybeare of Brislington near Bristol and later vicar of Axminster; and the practical and wealthy (at least for a while) Henry De la Beche who moved to Lyme Regis in 1812. All three were to carve out distinguished careers, but played such a significant local role that they have been christened 'the geological triumvirate of Lyme Regis'.[20]

They were part of what was then a developing network of geologists, palaeontologists, fossil collectors and museum curators principally centred around the recently-established Geological Society of London, but also with strong West Country connections, especially to Bristol and Bath where the formation of new philosophical societies and institutions was fuelled by a growing interest in science. Bath was where William Smith had pioneered his geological mapping, and extraordinary fossils completely new to science were being uncovered in Somerset, both along the coast and in the limestone quarries south of Bristol. But the focus of palaeontological discovery lay to the south, along the Dorset coast around Lyme Regis and Charmouth, and at its centre was the fourth member of Lyme's geological quartet, the professional fossil collector, Mary Anning.[21]

TWO

The Girl who Lived

As the eighteenth century gave way to the nineteenth, the England into which Mary Anning was born was experiencing huge social and economic changes. Farming practices such as the introduction of crop rotation, selective animal breeding, the enclosure of open fields and improved drainage had brought about an agricultural revolution. In Durweston, a small Dorset village near Blandford Forum where Mary Anning's mother was born, thirty smallholdings were reduced to two large farms. The old yeoman farming class became employed labourers on land they once had farmed. Hand-in-hand with such changes went the Industrial Revolution, whose reach extended from manufacturing, especially in textiles, to smelting, steam power, and the construction of roads and canals. To feed these new industries, and the factory towns springing up in the Midlands and North, the demand increased for raw materials – iron ore, limestone, coal. The population boomed, doubling between 1801 and the first census of 1841 from ten million to over twenty.

On top of all of this, war with France had dragged on from February 1793 and continued, with only brief breaks, until the defeat of Napoleon at Waterloo in June 1815. Lyme played only a minor role, provisioning the garrison on the Channel Islands. The only trade to flourish was smuggling, led by John Rattenbury, known as the 'Rob Roy of the West'. But few of the ordinary townsfolk could afford the contraband silks and brandy Rattenbury and his fellow smugglers landed by night on the beaches. The war had stifled imports. Much of the annual harvest was needed to feed the the army and navy, and wheat was often in short supply. Between 1792 and 1812, the price of bread almost tripled, leading to widespread unrest and rioting, especially following a poor harvest in 1799, the year Mary Anning was born.[1.]

One might think Dorset and Lyme Regis were provincial backwaters far removed from such events, but that was not the case. The poet William Wordsworth and his sister Dorothy were renting Racedown, a house 9 miles

(15 km) inland from Lyme in the mid 1790s. In a letter of 1797 he wrote 'The country people here are wretchedly poor'. An early version of 'The Ruined Cottage', written when at Racedown, describes 'Two blighting seasons when the fields were left / With half a harvest'. Famine struck Dorset in January 1800. Potato fields were dug over twice and anything green was eaten as a vegetable. A high tide stranded shoals of sprats on the shore at Lyme and entire families from the surrounding villages hungrily gathered them up. Wages were low, poverty widespread. In early March 1800, convinced local millers were hoarding grain in hope of the price rising still further, a mob descended upon several farms around Lyme and on what is probably now the Town Mill off Coombe Street in the town centre. The rest of the West Country was not immune. Rioting also broke out in Blandford Forum and in Bath and Bristol. One of the leaders of the Lyme riot was a carpenter and cabinet-maker called Richard Anning.[2]

THE ANNING FAMILY

Richard Anning had moved to Lyme from Sidbury, a small Devon village 12 miles west, where he was born in about 1766, the fourth of at least six children of William Anning, a yeoman, and Anne Flood, a labourer's daughter. Improvements in agricultural productivity were not matched by the living conditions of the rural workforce. Lyme Regis's reputation as a growing resort undoubtedly offered better opportunities for employment. By 8 August 1793, the date of his marriage in Blandford Forum to a labourer's daughter three years his senior, Mary (Molly) Moores, Richard is described as 'of Lyme Regis'.[3]

In December 1794, Molly gave birth, probably to twin daughters, Mary and Martha. Martha did not live long and was buried on 1 February 1795. A year later, in February 1796, Molly and Richard had a son, Joseph, who was baptised with his surviving elder sister, Mary, at Sidbury on 20 June 1796. Two other sons were to follow, Richard, born about 1797, and Henry, who died when still an infant and was buried on 17 April 1798.

Henry's death was not the last tragedy to be visited upon the family that year; young Mary met a dreadful end as it drew to a close. As the *Bath Chronicle* reported, on 27 December 1798:

> 'A child, four years of age of Mr. R. Anning, a cabinet maker of Lyme Regis, being left by the mother for about five minutes with her brother about three years of age, in a room where there was some shavings by their putting some into the fire, the girl's cloaths caught fire and she was so dreadfully burnt as to cause her death.'

So within four years the Annings had lost three of their five children. But

Molly was once again pregnant, and on 21 May 1799, some five months after the tragic death of her daughter, she gave birth to another, whom she again named Mary. Infant mortality rates were high – 36% of children failed to reach their ninth birthday – and it was not uncommon for a child to be given the name of a deceased sibling. This second Mary was to survive into adulthood and become the famous fossil collector. Further children followed: another Henry, Perceval and Elisabeth, but none lived longer than three months.[4]

Richard and his family were amongst a significant group of Dissenters in Lyme. Dorset's comparative isolation had enabled NonConformity to put down strong roots in the county, attracting tradesmen, small farmers and craftsmen – men like Richard Anning. They were also more political than their Anglican brethren, which perhaps explains Anning's role in the rioting over the price of bread. When the second Mary was baptised on 27 June 1799, it was not at Lyme's parish church of St Michael but at the Independent (Congregational) Chapel on nearby Coombe Street. The chapel had been built about forty years earlier, between 1750 and 1755 by its minister, the Reverend John Whitty who had come to Lyme in 1734. Whitty was not only the chapel's architect but also served as foreman, overseeing its construction and building the pulpit and galleries. Officiating at Mary's baptism was the Reverend John Crook from neighbouring Charmouth.[5]

Although no longer a place of worship, the chapel, the survivor of a fire of 1803 which destroyed the cloth factory and 42 houses, still stands in Coombe Street: appropriately, given the discoveries of the most famous person to be baptised in it, it is now a private fossil museum called Dinosaurland. It was here that the Anning children received a basic education, and learnt to read and write. Dissenter's Sunday Schools provided the only education for most working class families, too poor to afford day schooling. Through their childhood, the minister was the Reverend James Wheaton, born, like Richard Anning, in Sidbury, who took up the position in 1800. 'Amiable in spirit, faithful in friendship, and holy in conversation', the popular Wheaton saw his congregation swell.[6]

Mary Anning first went to the Chapel Sunday School when she was eight. Her brother Joseph, three years older, may have marked her start at school by giving her the 1801 volume of the *Theological Magazine and Review*. Hardly a children's book, it was a publication to which Wheaton occasionally contributed, and one which he may have originally given to Mary's parents, who in turn gave it to Joseph. Tom Goodhue, an American United Methodist pastor and author of several books and papers about Mary Anning, has considered the influence of NonConformity on her spiritual beliefs and suggests that her early formative reading and the sermons she heard in the chapel would have taught her to value her own self-worth and encouraged her to think for herself.[7]

Despite a limited education lasting only three or four years, Mary Anning's reading and writing skills were well above average for the first half of the nineteenth century. Illiteracy was commonplace, and only one in five were fully literate. Her spelling and punctuation were occasionally erratic but that was not unusual. As an adult she was able to conduct her business and personal correspondence and read and understand the scientific literature of geology.[8]

LIGHTNING STRIKES

Richard and Molly came close to losing their second Mary when she was fifteen months old. On the afternoon of Tuesday 19 August 1800, she was taken to a travelling equestrian circus in the Rack Field, an area of level ground above the east bank of the River Lim where the town's woollen mills stretched out their cloth on racks for drying. Mary was in the care of a woman called Elizabeth Haskings. It was a warm and sultry summer's day. In late afternoon the sky darkened. Torrential rain swept in over the coast. Carrying Mary, Mrs Haskings joined those crowding under a large elm tree for shelter. Suddenly, without warning, a deafening clap of thunder was followed by a bolt of lightning which struck the tree above them. Elizabeth Haskings, and two fifteen-year-old girls, Fanny Fowler and Martha Drower, were killed instantly. Mary was still in Elizabeth's arms. Many years later, her widower husband John described what happened next:

> 'The Child was taken from my wife's arms and carried to its parents in appearance dead but they was advised to put it in warm water and by so doing it soon recovered, and from that time it got better health … it may be said the death of the Nurse was the life of the Child.'[9]

Mary's remarkable survival instantly turned her into a local celebrity. As part of Lyme's folklore, the story of 'Lightning Mary' has been told and retold ever since, beginning in her lifetime with George Roberts' *The History of Lyme-Regis* (1823), who enriched it by adding that 'from a dull child, [she] became very intelligent'. While this might seem exaggeration, rare cases are known where trauma such as being struck by lightning or suffering a head injury has been linked with personality change and genius: sudden savant syndrome. We have no evidence, though, of Mary Anning being a child prodigy. An alternative explanation might lie in the care and attention she most likely received afterwards from her parents. Having lost one Mary to fire, and now nearly a second, Richard and Molly may have taken particular care of her upbringing, the attention making her more socially adept and self-confident.

The precise location of the Rack Field is uncertain, but is most likely to have been what is now an area of playing fields between Church Street and the River Lim. In the 1840s, when Lyme had two Parliamentary elections amidst claims of intimidation and vote-rigging, it was used for political demonstrations and celebrations. In June 1843, 'Mr Sand's celebrated Equestrian Troop' performed here, but whether Mary Anning attended is unrecorded. A century later it was the site of a tented camp of American soldiers preparing for D-Day. When houses were built in this part of the town in the late 1940s, the camp access road was extended and named Anning Road after Mary, although its link with site of the famous lightning strike seems coincidental.[10.]

We don't know where Richard Anning and his family were living at the time of Mary's birth, but from about 1808 they rented a house on the southern side of a small square, Cockmoile Square, adjacent to the Guildhall and prison in the lower, older part of the town, just to the east of the mouth of the River Lim. The whole area was redeveloped in the late nineteenth century, but in 1808 the square and surrounding streets were crowded with houses and businesses. An 1823 *Directory* lists a butcher, baker, grocer, hairdresser, straw-hat maker, dress-maker, milliner, confectioner, and several boot and shoe makers as well as what was by then the Annings' furniture-making and fossil selling business, all cheek by jowl.

This was working class Lyme, built of locally quarried Blue Lias, cob and thatch. Bridge Street, the principal route into the town centre from the east, was less than eight feet wide. Sherborne Lane, narrow and steep, was traditionally the home of Lyme's lace makers. The dye house and mills of the old cloth factory on Mill Green Street produced blanket cloth and a coarse narrow cloth known as 'Dorset dozens'. Hatches on the River Lim controlled the flow of the mill race. Following the fire of 1803 the factory was rebuilt, but still with a thatched roof, as were a row of cottages for the workers. In due course a second factory was built, and as late as 1823 they provided employment for 200, and this only two years after it was estimated that Lyme consisted of 480 families living in 401 houses. Some of these were the shops lining Broad Street, Lyme's main street and home of its principal inns. The higher uphill, the more substantial the house. The new middle class settlers built their villas overlooking the sea along Marine Parade or on the edge of town, complete with gardens, stabling, and uninterrupted views of the shipping dropping anchor near the Cobb.

An 1824 plan of Lyme shows the Guildhall and prison forming the northeastern side of the square with houses to the south and west and the square open to Cockmoile Street (now the eastern end of Bridge Street). Two houses formed the seaward, southern side of the square, a larger house to the east and a smaller, narrower house to the west, listed on the 1824 plan as 'A house, Fossil Shop & another Tenement'. The Annings occupied the larger house. Other buildings were situated between it and the sea.[11.]

A drawing of the property made in 1842 shows a building with three floors and a cellar, and a staircase rising from the cobbled square to a central door between two bay windows. It was a substantial house for a craftsman, but as well as the family home it was also Richard Anning's carpentry workshop, and later Mary's fossil shop.

In addition to his trade as a carpenter, Richard Anning established himself as a seller of fossils, one of the first in Lyme, according to George Roberts. He was obviously an unusual man, with an independent unconventional streak – characteristics inherited by his daughter. From a table set up in front of a basement workshop window, Anning, and later his widow and children, sold their fossil finds to visitors. The house was well-situated for access to the fossil-bearing rocks of Church Cliffs and Charmouth, and to the quarries which extracted limestone at Broad Ledge on the foreshore at the foot of Church Cliffs. A letter of July 1810 from a Bristol fossil collector, James Johnson, to a Wiltshire-based fossil collector and antiquarian, William Cunnington about to visit Lyme for his health, advised him to seek out

> 'a person at Lyme who collects for sale by the name of Anning. a Cabbinet Maker and I believe as men are may be depended upon. I would advise you calling upon him at Lime and he will point out to you comfortable lodgings there or at Charmouth which is only two miles distant ... a Mr Carpenter surgeon called upon me ... he is a collector at Lyme and a resident. do call upon him.'[12.]

George Roberts also tells us that Richard Anning,

> 'being a dissenter, went on the beach on Good Friday and church holidays. If he found any thing worth purchasing, it was usually exposed in front of his shop, near the prison, on a little table. He occasionally went out on other days, at which his wife was very angry, and was wont to ridicule his pursuit of such things.'[13.]

Mary Anning later attributed all that she knew about fossils to her father, so clearly she accompanied him on some of his fossil-hunting excursions.

CURIOS AND CURIOSITIES

By the end of the first decade of the nineteenth century, Lyme and Charmouth had become well-known for their *curios* or *curiosities* – fossils. Collecting and selling probably first started at Charmouth as it had the advantage of being on the London to Exeter stage coach route. In 1788, the Reverend Stebbing Shaw, a Derbyshire rector, pausing in Charmouth whilst the horses were fed and watered, met a local labourer, William Lloyd. In his book, *A tour to the West of England in 1788*, published the following year, Shaw recalled:

> 'Meeting William Loyd, a labourer, we were induced to accompany him to see his collection of the most curious fossil world. His cottage affording no convenience for this purpose, they are displayed in the open garden; those who are desirous of viewing such wonderful operations of nature, may here satisfy their curiosity by only deviating a few yards from the road; and those who are desirous of adding to their collection for grottos, chimney-pieces, &c. may here find materials on the lowest terms.'[14.]

Lloyd was clever enough to display and market his fossils not merely as souvenirs, but as decorative embellishments. It was a trend that caught on. Many a Dorset cottage still boasts an ammonite set into a wall or displayed in a garden. But he may have had competition, as George Roberts wrote later:

> 'About the time that watering-places arose out of the old towns, fossils, called curiosities, became an object that attracted the visitors. One Lock, whom Dr. Maton calls *curi-man*, used to offer to the coach passengers at Charmouth specimens of *verterberries, mushrooms, ladies' fingers, John Dories,* and *cornumoniuses.*'[15.]

The 'Curi-man' [Curiosity-man], or 'Captain Cury' was William Lock from whom Dr William George Maton bought 'some fine specimens of chalcedony' to add to his collection while touring the West Country in 1794. Not all thought so highly of Lock. The Bristol fossil collector James Johnson, a regular visitor to Charmouth since 1789, in his letter to William Cunnington, suggested that he

> 'should walk to Charmouth and ask a confounded rogue of the name of Lock to call upon you ... upon first sight give him a Grog or a Pint this will buy him to your interest and all Crocodiles he may meet with will most assuredly by offer'd you first, you must then agree with him for he is poor and will sell within one hour after the article is found.'[16.]

By contrast, wrote Johnson, Richard Anning could be 'depended upon', whilst Roberts noted that 'Richard Anning sold similar objects at Lyme, having learnt how to find such things by accompanying a Mr South, who is believed to have been employed by the Duchess of Portland'. Hugh Torrens has identified Anning's mentor as John South who was born in Wiltshire, but by 1798 was living in Lyme, a move he may have made specifically to collect fossils. There he met James Johnson who occasionally accompanied him on expeditions on the foreshore.[17.]

Another of Richard Anning's fossil-collecting companions was John Crookshanks, a London coal merchant who had moved to Lyme and lived on a modest annual allowance. According to Roberts in 1834, Crookshanks

> 'was the first collector of *curiosities* at Lyme ... He went upon the shore on search of them, with a long pole like a garden-hoe, and had found many vertebrae, called here *verterberries*, and *fragments of jaws*, &c.' Richard Anning used occasionally to accompany Mr Crookshanks. The fragments of fossils they found were considered to be, and so named, *bones of crocodiles' backs and jaws, ladies' fingers, John Dories, salmon* &c.'

Alas, when his allowance was discontinued in 1802, Lyme's first collector committed suicide by jumping from Gun Cliff into the sea.[18.]

The term 'collector' is ambiguous; it might mean a John Crookshanks actively searching the rocks and cliff falls for fossils, or someone who buys specimens found by others. Many collections were built by a combination of the two, either as a hobby or for scientific study. For others, collecting was fashionable, a rare or spectacular fossil becoming a show-piece in a gentlemanly 'cabinet of curiosities'. Then there were collectors like the Annings, commercial fossil hunters whose finds were their stock for sale.[19.]

AN ERODING COAST

The fossils which Roberts refers to by their common local names include what we now recognise as the vertebrae of the marine reptiles, ichthyosaurs and plesiosaurs ('verterberries'); ichthyosaur jaws and skulls ('crocodiles'); the bullet-shaped, internal shell of the extinct squid-like molluscs, belemnites ('ladies' fingers'); the coiled, spiral shells of ammonites ('cornumoniuses'); and the fossil fish *Dapedium* ('John Dories'). That these fossils were readily found, and widely known about, is the result not only of their abundance in the Lias rocks around Lyme and Charmouth, but also of coastal quarrying and cliff erosion continually exposing new material.

Above Church Cliffs is Lyme's parish church, St Michael the Archangel. Church Street, on which St Michael's is situated, continues uphill to become Charmouth Road and defines the eastern edge of the town. That St Michael's still stands is in itself remarkable. The churchyard extends a short distance beyond the building and ends, abruptly, at the top of Church Cliffs, the source of much fine fossil material in the early part of the nineteenth century. The cliff here was then eroding at a rate of about a metre a year through landslipping. Between about 1750 and 1810 much of it disappeared, taking with it the town's first Walk, where Lyme's smarter visitors liked to stroll, along with the lane leading to Charmouth.

George Roberts noted in 1823 that 'an old woman, living in 1755, said she remembered, when an apprentice, milking her master's cows three fields farther towards the south than the land now extends behind the Cobb, which must have been nearly parallel with the pier-heard at that time'.[20]

The instability of the cliffs was not only a concern to farmers fearing the loss of good grazing, but also house owners. Lyme physician and fossil collector, Dr Thomas Carpenter, took an interest in the rate of erosion of Church Cliffs. According to Mary Anning, in an 1829 letter to the geologist Charles Lyell, in 1800 Dr Carpenter

> 'paced the Church Cliffs from a certain hedge and that it was thirty paces to the edge of the Cliff ... it is fallen away from that hedge so as to leave about one yard between the hedge and the edge of the Cliff having lost 29 yards in as many years.'[21]

RICHARD ANNING MEETS DE LUC

In September 1805, the Swiss geologist Jean André De Luc made the first of two visits to Lyme. He travelled from Bridport by coach to Charmouth and then walked along the foreshore to Lyme where he noted the presence of large ammonites:

> 'there are people who watch the crumbling down of the cliffs in winter, and search among the fallen materials for the fossils contained in them, and especially for these *cornua Ammonis*, which make a very ornamental appearance in collections.'[22]

Once in Lyme he

> 'made the acquaintance here with one of the inhabitants, who knows this coast very well, because he visits it, from time to time, in search of fossils, which he sells to the strangers who resort hither; Lyme being a place much frequented for sea-bathing. At his house I saw some find [fine] *cornua Ammonis*, sawn through the middle, and various other marine fossils.'

This was almost certainly Richard Anning, probably the only person then selling fossils in Lyme. De Luc's comment that he saw cut sections of ammonites, tells us that Anning was not just collecting, but using his cabinet-maker's tools to prepare the specimens to render them more attractive and saleable – skills he passed on to his son and daughter.

De Luc was interested in the erosion he had noted on his walk and thought it a potential danger to the town. The two men walked up to the church and Anning showed De Luc the crumbling cliffs encroaching upon the churchyard. He was convinced it was only a matter of time before the church collapsed, followed in due course by the houses on that side of the town.

De Luc recognised that part of the problem was the flow of water and springs within the cliffs. He suggested a remedy involving boring into the cliff to intersect the springs and the construction of a strong sea-wall at the base of the cliff. It would, he added, be expensive. In a comment that still resonates today, De Luc noted that 'the present generation does not believe itself in danger; and generations consider but little the interests of those which are to succeed them'. Two hundred years would pass before De Luc's recommendations were finally implemented.[23]

In the course of their conversation, Richard Anning described being caught up and almost killed in a landslide here. Walking alone along the cliff path close to some recently-developed fissures, the ground suddenly gave way beneath his feet. In the nick of time, clods of earth falling around him, he leapt up onto the higher stable part of the cliff and hauled himself to safety.

This may not have been Richard Anning's first fall on these cliffs, and it was not to be the last. Three years later, in early 1808, he was walking to Charmouth after dark and 'having diverged from the path in a field on the summit of the hill, he went over [the cliff] where the new road now is, on the west end of the deep cutting.' This time he did not escape so lightly. The fall was long, his injuries serious. Unable to work, the family faced hardship. Richard contracted tuberculosis and he died in November 1810, at the age of 44. He was survived by his wife, Molly, his son, Joseph and daughter, Mary and, it seems, by at least one other son, Richard, who was also dead within a year.[24]

Richard's death had left the family £120 in debt. For the next few years they were dependent on charity, receiving aid from the overseers of the Parish Poor from 1811 until at least 1816. The Annings were not long-established in Lyme, and had no local family network to turn to for support. Dependency on parish relief often meant destitution, and the shame and fear of being forced into the Poor House. Mary Anning was eleven when her father died, on the cusp of adolescence, and there can be little doubt that the years of poverty left psychological scars that never completely healed. Worries about money and earning a living never left her. Despite her later fame, the friendships she forged, she always remained an outsider. The only legacy Richard left his children was all they had learnt at his side, either amidst the shavings of his carpenter's workshop or scrabbling amidst the mud and rubble at the foot of the cliffs. Joseph became a cabinet maker and upholsterer, whilst a chance find set his sister on the career by which she is remembered.

THREE

The First Ichthyosaurs

AT LEAST SOME OF THE LEGENDS that attached themselves to Mary Anning stem from the pen of George Roberts. Roberts was born in Lyme in 1804, so was five years younger. Historian, schoolmaster, the magpie collector of every anecdote he could trace about his native town and the author in 1823 of its first history, he was also a benevolent witness to Mary Anning's changing fortunes, taking every opportunity to publicise her finds.

Shortly after her father's death, wrote Roberts later, Mary Anning went down to the beach to look for fossils and found an ammonite. On her way home, 'a lady in the street, seeing the fossil in her hand, offered half-a-crown for it, which she accepted, and from that moment fully determined to go down "upon beach" again.'[1.]

The same incident is recorded in the journal of Anna Maria Pinney, the eighteen year-old member of a well-known Dorset family. Pinney befriended Mary Anning in the early 1830s and, describing an October day in 1831 spent 'fossilizing' with her, wrote

> 'She told me at ten years old her father died, and her mother being in great distress did not attend to her, she wandered along the seashore, and picked up a fossil, which she showed to a lady whom she met – this lady was a fossilist and was so much pleased that she gave her ½ a crown.'[2.]

Another visitor to Lyme, a fourteen year-old girl, Frances Bell, recorded the same story from Anning in 1824, and places it to a specific date:

> 'The day after her father's funeral, she accidentally, or rather, as she piously observes, providentially picked up a fossil, while rambling on the shore, which she sold for two shillings and sixpence: this give her a stimulus.'[3.]

In these three accounts, all retold long after the event, Mary Anning links the beginning of her fossil collecting career to the greatest trauma of her childhood: the loss of her father. The last two entries as good as imply that

Mary was neglected by her mother Molly, still in shock and mourning the loss of her husband, and took to wandering the beach, not necessarily in search of fossils. The finding of the ammonite seems to have been fortuitous (or providential). Its sale to a fellow fossilist was the catalyst for Mary realising that she could earn her living collecting and selling fossils. Two shillings and sixpence was the approximate equivalent of two days wages as a labourer. Molly Anning had been dismissive of her husband's fossil collecting, but her daughter's find and its subsequent sale helped win her round: the income from selling fossils could at least help alleviate their poverty.

A year later, in the autumn of 1811, Joseph was looking for fossils on the beach between Church Cliffs and Charmouth when he made a remarkable discovery. Part of the cliff had collapsed and brought down a pile of rock. Exposed within it were the bones of a fossil skull over four feet long. Once the surrounding rock was removed, it was apparent that the huge skull was unlike any known animal, although resembling mostly a crocodile. It had a huge eye socket containing a ring of bony plates, and a long, tapering snout with jaws packed with sharp, conical teeth. A year later, in early November 1812, the rest of the skeleton, 17 feet (5 m) long, was recovered by his sister. It took scientists years to identify their find. By 1820 it was recognised as a new kind of extinct marine reptile and was named *Ichthyosaurus*, 'fish-lizard'.

The discovery was described over thirty years later in a letter by Joseph's son, Charles Churchill Anning:

> 'After the death of Richd. Anning Mary Anning's Brother went down to look for Fossils and he then with the help of two men he hired to assist him dug a large head of a Crocodile as it was supposed. that part of the Beach was not again opened or uncovered for nearly 12 months afterwards when Mary Anning discovered the remainder having been shown by her Brother the spot where he dug out the Head, and requested by him to look out for the remains when the Beach was again laid open after a storm. This crocodile so it was considered was traced back as it lay by her and dug out by men she hired.'[4.]

Anna Maria Pinney's journal also tells the story that after receiving her half crown, Mary 'persisted in searching among the rocks and the following year discovered the first Ichthyosaurus that was found', while Roberts recounts that

> 'four months after, in the year 1811, [she] saw among the ledges a projecting bone of some animal. This crocodile (so it was considered) was traced as it lay by her, and dug out by men she hired.'

There is no mention of Joseph's involvement. That the specimen was recovered in two stages, a year apart, is reported in the first scientific paper to describe the fossil, published in 1814. We don't know why Joseph passed the search for the skeleton to his sister, but by then he had embarked on his apprenticeship as an upholsterer, whilst she had more time to spare.

The date of what seems to be the retrieval of the remainder of the skeleton is given in a newspaper report on 9 November 1812:

> 'A few days ago, immediately after the late high tide, was discovered, under the cliffs between Lyme Regis and Charmouth, the complete petrifaction of a crocodile, 17 feet in length, in a very perfect state. It was dug out of the cliff nearly on a level with the sea, about 100 feet below the surface of the earth.'[5]

This suggests that what was excavated was a complete skeleton, the 'complete petrifaction', and makes no mention that the skull was found a year before, an event that seems to have gone unreported. Some contemporary newspapers are even more confusing, seemingly combining the discovery with its presentation to a London museum:

> 'Dorsetshire. A part of the sea cliffs on the coast, near Lime, lately fell down, after a violent storm, and discovered the fossil remains of an enormous crocodile, in a state of perfection not before found. This extremely valuable relic was discovered on the estate of H.H. Henley, esq. who has liberally presented it to the London Museum of Natural History.'[6]

It seems, then, that the skull was discovered by Joseph Anning in the autumn of 1811, a year after his father's death, and the rest of the skeleton recovered in early November 1812 by his sister. On both occasions, according to Charles Churchill Anning, men were hired to help dig the specimen out. These were probably quarrymen working the limestone ledges on the shore at Broad Ledge who had the tools and expertise. It is possible to imagine Joseph Anning, as a fifteen year old, hiring them, but it seems less credible that a thirteen year old girl might do so. It is much more likely that it was Molly, their mother, who negotiated any charges. However, Molly was still in debt and unlikely to be able to pay them. But Lyme was a small town, and many knew of the family's predicament and Mary's miraculous survival from the lightning strike twelve years earlier. It was also an astonishing discovery, unlike anything anyone had seen before, and it seems probable that the quarrymen were only too glad to help.[7]

FOSSIL 'CROCODILES'

We know that the Lias rocks of the Dorset coast contain fossils not just of marine shells such as bivalves, gastropods, ammonites and belemnites, but also of back-boned animals like fish and two different kinds of marine reptiles, ichthyosaurs and the related plesiosaurs. While most of the invertebrate shells are readily recognisable for what they are, mainly due to their resemblance to their modern relatives, and fossil fish can often be well preserved so that identification is straightforward (hence the name 'John Dories' given to the fossil fish *Dapedium*), identifying the marine reptiles is not so easy. More often than not, the bones are dispersed and found isolated; complete, intact specimens are rare; and of course, there are no modern marine reptiles for comparison.

Fossil bones had long been known from the Lias rocks of Lyme and elsewhere in England and Wales. The first, thought to be 'great bones of fishes found in the earth' were illustrated as early as 1605 and can be recognised today from seventeenth century engravings as the vertebrae of plesiosaurs. The earliest known illustration of ichthyosaur vertebrae, from Purton Passage in Gloucestershire, dates to 1699, and again these were thought to belong to fish. Their vertebrae are distinctively biconcave, so that a section through one has an hour-glass shape in a similar fashion to fish vertebrae. In Dorset, collectors such as William Lock, as well as Mary's own father, Richard, and no doubt Mary herself, were certainly familiar with the isolated vertebrae which were known to them, and sold, as 'verterberries'.

Throughout the eighteenth century, more fossil material was discovered, including some partial skeletons. As it became apparent that they were not fish, they came to be called 'crocodiles', the known living animal whose bones and teeth the fossils most closely resembled. Some, especially those from the Lias rocks of Whitby in Yorkshire, were indeed crocodiles, but many, we now believe, were ichthyosaurs, although they would not be recognised as new or named until 1817. Many of these fossil 'crocodiles' found their way into private collections, often of clergymen, and only later did some go to public museums. Most were not complete enough to show the main features that differed from crocodiles, although one nine feet (3 m) long ichthyosaur discovered in about 1805 near Bath belonging to the Reverend Peter Hawker had seventy vertebrae up to 9 inches (22 cm) across and a skull three feet (1 m) long with 120 teeth in its jaws. These details of what became known as 'Hawker's crocodile' were published in the *Gentleman's Magazine* in January 1807. Its significance was not recognised until after the Anning specimen had come to the attention of scientists. So while the specimen found by the Annings was not the first ichthyosaur skeleton to be discovered, it was the first specimen to be scientifically studied and described, and along with other material found subsequently, given the name *Ichthyosaurus*.[8.]

The specimen found by the Annings was purchased from them for £23 by

Henry Hoste Henley of Sandringham Hall, Norfolk (the estate later bought by Edward VII and now belonging to the Queen) who was also Lord of the Manor of Colway on the edge of Lyme. As it had been found on his property between Lyme and Charmouth, the Annings may not have had much choice in the matter. Be that as it may, the £23 represented almost a fifth of Molly's debt and the equivalent of about six months wages for her late husband. If she still had any doubts about the possibilities offered by selling fossils, this must surely have dispelled them.[9]

The press reports of the discovery and sale, and the subsequent scientific interest in the specimen helped establish the family's reputation as fossil collectors. They were certainly aware of the significance of their find, inspiring a search for more. Young Mary's interest in fossils was encouraged by a Mrs Stock who lived in the Great House on Broad Street. Charlotte Stock was the wife of a Bristol physician, Dr John Edmonds Stock. She employed Mary to run errands for her, and when she was fourteen 'gave her the first book on Geology she ever read'. We have no record of what book this was, but it may have been Robert Bakewell's *Introduction to Geology*, first published in 1813.[10]

By 1820, Molly, Joseph and Mary were set up in business as the 'Fossil Shop'. Although we have scant information about the family's finds in the wake of the discovery of the first ichthyosaur, it seems certain they provided additional specimens to established collectors like James Johnson and Henry De la Beche and to other clients. These further specimens helped elucidate the true nature of the animal the Annings had found. De la Beche played an active role in this research and would have kept the Annings informed of the discussions and arguments amongst the scientists in London studying their specimens.[11]

THE CROCODILE GOES TO LONDON

Henley presented the specimen in 1813 to a new museum, the Egyptian Hall, which had recently been built and opened on Piccadilly in London by the collector and antiquarian William Bullock. Listed in Bullock's published catalogue as 'Fossil of the Head of an Animal, resembling a Crocodile of enormous size. This wonderful object was found in Dorsetshire', the specimen was displayed at the Egyptian Hall until 1819. That year Bullock decided to auction off the contents of his museum in a sale which lasted twenty-six days from 29 April to 11 June and attracted buyers from museums across Europe. The specimen sold as lot 100, 'Crocodile in a Fossil State', on the ninth day of the sale and was purchased by Charles Konig, Keeper of Natural History at the British Museum for £47 5s, double the amount that Henley had paid the Annings. According to Charles Churchill Anning, Henley was much hurt at its having been sold with the other articles at the breaking up of that collection.[12]

While at Bullock's Museum, the Anning specimen had come to the attention of Sir Everard Home, Professor of Anatomy and Surgery at the Royal College of Surgeons. He described and illustrated it in a paper published in 1814 in the *Philosophical Transactions of the Royal Society*. In this he tells us that the specimen was found

> 'on the estate of Henry Host Henley, Esq. between Lyme and Charmouth, in Dorsetshire, in a cliff thirty or forty feet above the level of the sea-shore. It had been thrown down by the breaking off of a part of the cliff and buried in the sand upon the shore, to the depth of nearly two feet. The skull was dug out in 1812, the other parts in the following year, at a distance of some feet.'[13]

It is at this point that the specimen begins to lose its association with the Annings. Only Henley on whose land it was found and who gave it to Bullock's Museum is mentioned. By 1835, the connection of that first ichthyosaur with the Annings was forgotten even by a friend of Mary's, the Oxford geologist William Buckland. In a letter to Charles Konig at the British Museum, Buckland asks Konig about several ichthyosaur specimens in the museum. One is a specimen illustrated by a collector called Thomas Hawkins, while the other is 'the large Ichthyosaurus . . . described by Sir E. Home from Lyme & obtained from Bullock'. Only loyal George Roberts made certain the Annings received their due, giving a full account of its discovery and its eventual acquisition by the British Museum in the 1834 second edition of his *The History and Antiquities of Lyme Regis*.[14]

Although Home has the dates a year later, his account of the two-stage discovery agrees with that given by Charles Churchill Anning and, broadly, with the contemporary newspaper report regarding where it was found. The Henley family property, according to the 1841 tithe map, lay beyond Church Cliffs. If this was the case in 1811, the sand-covered specimen must have been found somewhere between East Cliff, about 200 metres east of Church Cliffs, and Black Ven.

Home's paper was accompanied by four finely-drawn plates prepared by William Clift, Conservator at the Royal College of Surgeons. One shows the skull, another a block of rock with part of the vertebral column, ribs, left shoulder and some front limb bones. The other plates illustrate a block of rock with some ribs, and a cut section through a group of three vertebrae. Not pictured, but mentioned by Home in the text are 'the number of vertebrae collected, which appear to have formed one connected chain, is sixty, and when these are placed in a line, the skeleton measures about seventeen feet'. This measurement is consistent with that in the November 1812 newspaper report.[15.] Home puzzled over the animal's identity, considering some bones to be like those of a crocodile, some like a turtle, and others like a fish. Ultimately flummoxed, he played safe by inclining to the view that it was a fish, but not 'wholly a fish'.

Two years later, Home examined additional fossil specimens, some belonging to William Buckland and a complete, but single, limb in the collection of James Johnson which had been discovered in the summer of 1814 by Mary Anning at Black Ven. These new finds confirmed Home's belief that the animal was indeed a fish, but an extinct one.

Two more years passed by which time Home had seen further specimens belonging to the Reverend Peter Hawker in Gloucestershire and Dr Carpenter and Henry De la Beche in Lyme, from whom he had also been sent a drawing of a skull. All of this made Home change tack. In a paper published in 1818 he stated that although it swam like a fish it wasn't one after all, but breathed air and had some similarities to the *Ornithorhynchus* – the duck-billed platypus of Australia.

Others joined the debate into the exact identity of the animal being found in the Lias rocks. James Johnson, with thirty years' experience of collecting fossils on the Dorset coast, put forward his own theory. As the *Annals of Philosophy* reported in 1815:

> On Tuesday, December 6th [1814], a paper by James Johnson, Esq. was read [at the Linnean Society in London], giving an account of some fossil bones found in the cliff near Lyme, Dorsetshire. This cliff abounds in Belemnites, nautili, and the remains of other sea animals. The bones in question have been supposed to belong to the crocodile; but Mr Johnson gave his reasons for considering that opinion as ill founded. He thinks they constitute the bones of a new and unknown species of amphibious animal. He is of the opinion that the animals whose remains are found here lived and died upon the spot.'[16.]

Johnson's suggestion that the animal was air-breathing was made four years before Everard Home came to the same conclusion.

In March and April 1819, in his fourth and fifth papers to the Royal Society, Home finally gave a name to the animal. With new material from De la Beche and a spectacular complete specimen to examine, probably another Anning discovery, showing that it had four paddle-shaped limbs, Home concluded that it was closest to lizards and salamanders so he named the fossil *Proteosaurus*. Proteus was a mythological Greek god of seas and rivers who gave his name to a genus of salamanders first described in 1768, while *sauros* is Greek for a lizard or reptile. (Hence familiar dinosaur names like *Brontosaurus*).

Home's confusion is understandable. The notion of creatures becoming extinct was a novel one, and only beginning to be understood and accepted. When a new species of fossil animal was discovered, it had to be fitted into the spectrum of known life, hence the first identification of ichthyosaurs as crocodiles. All organisms were considered to be linked to one another in a continuous chain, an ancient idea called the 'Great Chain of Being', into which fossils were fitted by identifying their similarities to known animals. As increasing numbers of complete specimens came to light, the 'Great Chain of Being' looked increasingly vulnerable. Some skeletons bore little resemblance to any known animal still living on the planet.

Some philosophers and scientists argued that these unknown animals were still alive and awaited discovery in some remote corner of the earth, usually jungle, but for others a new explanation had to be found. In 1856, George Roberts recalled this time when 'Fossilisers, in later years called geologists … were almost universally viewed in a bad light, as infidels and perverters of the Scripture.'[17] The possibility that not every animal God had created still existed was contrary to accepted Christian thought, causing particular difficulties for the Anglican clergy in the late eighteenth and early nineteenth centuries, especially as some were at the forefront of the new science of geology. The conundrum lay in reconciling Biblical teachings with the fossil evidence coming out of the ground.

The idea of extinction had been mooted seriously just fifteen years before Joseph Anning found the fossil skull at Lyme. George Cuvier was then a young palaeontologist and comparative anatomist at the Natural History Museum in Paris who was studying drawings and specimens of fossil bones of large land animals found in South and North America and in Siberia. While the bones had similarities to those of some living animals, they were different enough to defy classification. Some from South America, he realised, were from a giant sloth, larger than anything living. Others found in North America were from a kind of elephant, but one different from Indian and African elephants. Elephant bones from Siberia had thick fur still attached, again unlike living elephants.

Cuvier thought it unlikely that such large animals were still alive – somebody, somewhere would have seen or heard about them. Closer to home for Cuvier, in the vicinity of Paris, fossil bones were being discovered from elephants and other mammals that were clearly no longer living there. These fossil animals were extinct. Discoveries like those of the Annings were proof that fossil-bearing rocks contained animals unlike anything now alive. And these were not from far-flung places like Siberia or the Americas, but from provincial France and rural England.[18]

In 1813 Cuvier attempted to explain extinction by invoking periodic disasters, such as inundation by flooding, to wipe out some animals, allowing others to move in. The belief that the earth and life on it had been shaped by such sudden catastrophic events became widespread in the early nineteenth century, and although Cuvier avoided soliciting divine intervention as an explanation, others saw the Biblical Flood as a perfect example of this catastrophism.

It was particularly attractive to the Anglican clergy-geologists, for whom the Deluge explained the shape of the landscape and the widespread surface deposits of gravel and boulders. These deposits were termed diluvium, and William Buckland became a leading advocate, calling his 1823 book describing the fossils and deposits found in caves such as Kirkdale Cavern in Yorkshire, *Reliquiae Diluvianae*, (*The Relics of the Flood*). By the late 1820s, advances in the understanding of geology made the idea of a single large flood implausible. In the 1840s the idea was abandoned by Buckland when the Swiss geologist Louis Agassiz proved that most of the deposits and landscape features in Britain attributed to the Flood were the result of the action of glaciers.

Cuvier's catastrophism and its need for abrupt changes was dealt a blow in the early 1830s by the publication of *Principles of Geology* by the Scottish geologist Charles Lyell. Lyell argued that the processes acting slowly to shape the earth today, like the movement of sediment by rivers or the erosion of coastal cliffs by the sea, had always operated in the same way, gradually modifying the planet throughout geological time. Time was the key; all that was needed was lots of it, literally all the time in the world. In this, Lyell was building on the ideas of another, earlier, Scottish geologist, James Hutton who had suggested in 1788 that today's earth processes have been in operation throughout earth history slowly changing the planet and that geological time was unimaginably vast. 'We find,' Hutton wrote, 'no vestige of a beginning, – no prospect of an end.'[19]

Georges Cuvier's fossil studies led him to speculate that before mammals were the dominant forms of life, there had been an age when reptiles ruled. The fossils being found by the Annings, and later by William Buckland, Gideon Mantell and others, were providing the evidence of just such an Age of Reptiles. With their discoveries, the Annings were opening a door to the lost world of an ancient Dorset, a world inhabited by strange marine reptiles whose remains were forcing early nineteenth century Christians to reconsider the story of Creation and the age of the Earth.[20]

FOUR
Friends and Neighbours

FROM HOME'S SERIES OF PAPERS, it is clear that additional specimens were found between about 1811 and 1819 in Lyme. Similar fossils were also coming to light in Somerset and Gloucestershire, often ending up in the hands of West Country collectors. One who took a particular interest in the new animal, and provided Home with specimens and drawings, was Henry De la Beche. He was one of a group of geologists with close links to Lyme who became friends with Mary Anning. Thanks to them, her major discoveries were brought to scientific attention and she in turn learnt of the significance of what she was finding. Despite her lack of education, she read their published papers and books and, over time, developed a knowledge of Lias fossils at least the equal of her academic male colleagues at Oxford and Cambridge. She also became part of a circle of female friends, whose influence and encouragement did much to ease her sense of isolation from the working class Lyme of her birth: Elizabeth, Margaret and Mary Philpot, and Charlotte Stock in Lyme; Sarah Kennaway in Charmouth; as well as by Mary Buckland in Oxford and Charlotte Murchison in London. Most of these women were between ten and twenty-five years older than Mary Anning: their common ground was an interest in fossils. Like her male friends, these women were middle class and of comfortable means. We know almost nothing of any friendships she may have had with her neighbours in Cockmoile Square or her contemporaries at Sunday school.

HENRY DE LA BECHE

Henry De la Beche had moved to Lyme in 1812, and may have been in the town when Mary Anning extracted her ichthyosaur's skeleton. He was born in London, the son of a cavalry officer. His father had inherited a slave-worked sugar estate in Jamaica and, on a family visit in 1801, had died there. Henry was sixteen when came to Lyme with his, still young but by then twice-widowed, mother, Elizabeth, and her third husband, William Huddle

Aveline. While a schoolboy near Bristol Henry developed an interest in fossils, finding belemnites and ammonites – 'thunderbolts and snakestones' as they were often then called – on the banks of the River Avon. In 1810 De la Beche was sent to the Royal Military College at Great Marlow, which he enjoyed much more than he had school. However, he had spent much time on his own as a child, especially during a long convalescence from scarlet fever. Characteristics such as a sense of humour, boundless energy, a questioning mind and a dislike of authority did not endear him to the army. After eighteen months he was dismissed for insubordination, joining his mother and new stepfather in Dawlish in south Devon in 1811.

A year later they moved east to Dorset, first to Charmouth and then to Lyme. Their motive remains unclear, but it may be more than coincidence that Sidmouth, Ilfracombe and Dawlish, where De la Beche and his mother lived at various times were, like Lyme, all developing seaside resorts. Nor do we know where they lived when they first came to Lyme, but in 1814, De la Beche's stepfather took out a lease on Woodville, a villa near the top of Silver Street.[1.]

With his interest in fossils, it surely cannot have been long before De la Beche made the acquaintance of the Annings, especially with the excitement following the recovery of the ichthyosaur skeleton in November 1812. He was the same age as Joseph Anning, and three years older than Mary. Over the course of the next few decades he and Mary became close friends, despite their quite different stations in life; she and her family on parish poor relief, he the heir to a Jamaican sugar estate.[2.]

The young De la Beche's geological interests were supported by his stepfather and his friends in Lyme, especially Dr Thomas Carpenter and George Holland, both collectors of fossils. As the local physician, Carpenter knew the Annings and had attended those killed in the lightning strike of August 1800.[3.]

In February 1817, at the age of twenty-one, De la Beche gained his inheritance, held in trust for him since his father's death sixteen years before. He travelled to London and joined the Geological Society, and back in Lyme bought himself a yacht which he proudly sketched in one of his journals. In this he may have been encouraged by George Holland who also owned a Lyme-based schooner. De la Beche's fossil collection grew through purchases and his own collecting, sometimes accompanied by Mary Anning. The Wiltshire fossil collector, Miss Etheldred Benett wrote of him in 1818, 'Mr De la Beche is indefatigable ... he pays the collectors so high for whatever he finds himself as well as for what they find that he will be sure to secure all that is worth having.' De la Beche was particularly generous to the Annings, buying specimens from them, even when he had found them himself on joint expeditions to the beach.[4]

De la Beche's collecting gradually found focus, becoming concentrated on ichthyosaurs. While Sir Everard Home was deliberating on the affinities of the specimen found by the Annings, Charles Konig at the British Museum, coined the name '*Ichthyosaurus*' or 'fish-lizard' for the new animal, in 1817. The name was soon adopted and used by De la Beche in his private journals, describing his own specimens and others he had examined belonging to other collectors.[5]

His journals also mention that he had been 'frequently present at the taking out of the skeletons of the animals', much of this work undoubtedly by the Annings. So within seven or eight years of the discovery of that first specimen ichthyosaurs were no longer the rarity new to science that they once were, especially now that collectors knew what to look for. De la Beche himself found two within ten days in February 1819, one between Lyme and Charmouth, the other to the west of Lyme.

The interest sparked by the Annings' initial find led not only to more discoveries in the rocks at Lyme but also to the recognition, in private collections around the West Country, of ichthyosaur specimens that had been collected previously. In April and May 1819, De la Beche toured a number of these collections, examining the specimens, some previously assumed to be crocodiles. Amongst the collectors he visited were James Johnson at Hotwells in Bristol, the Reverend Peter Hawker in Stroud, and Charles Wilkinson in Bath before heading to Oxford to see William Buckland's collection and on to London in time for Bullock's auction sale. While in London he visited Everard Home at the Royal College of Surgeons, taking with him one of his own ichthyosaur specimens from Lyme. Step by step, each new specimen discovered or examined was helping build up a picture of these strange dolphin-shaped animals.[6]

By early January 1819 De la Beche's research had shown that there was not just one kind of ichthyosaur present in the Lias rocks, but at least three, and that some must have been huge. One fossil skull in the collection of James Johnson in Bristol had an eye socket 14 inches (37 cm) across and must have belonged to an animal almost 25 feet (7.5 m) long. In his own collection, De la Beche had a vertebra almost 8 inches (20 cm) across. The Lias sea must have been a terrifying place with these enormous, sharp-toothed monsters patrolling its waters.

De la Beche's researches were not restricted to the local fossils, but extended to the rock strata of the Devon and Dorset coast. In March 1819, he presented his first paper at the Geological Society on the geology of the coast between

Bridport and Babbacombe Bay. In it he drew attention to the occurrence of ichthyosaurs in the Dorset Lias and pointed out that 'the remains of this animal are by no means rare; they are principally discovered at Black Ven, and most commonly in the slaty or marly part of the lias.' He goes on to state that he had recognised three species of *Ichthyosaurus*. To the smaller, more common specimens he gave the name *Ichthyosaurus communis*, occasionally called *Ichthyosaurus vulgaris* in some reports; those with a long, slender snout he named *Ichthyosaurus tenuirostris*; and the largest specimens, which had slightly flattened teeth, he called *Ichthyosaurus platyodon*. In a later paper he added a fourth species, *Ichthyosaurus intermedius*.[7]

De la Beche's researches in southwest England continued through the early 1830s when he began making a geological map of Devon. It was then that the income from his Jamaican estate failed due to unrest on the island connected with the abolition of slavery and competition from sugar producers in Cuba. He successfully approached the government for the funding he needed to complete his mapping of Devon, and in 1835 continued his work into Cornwall as Geologist to the Ordnance Trigonometrical Survey. From this appointment grew today's British Geological Survey. De la Beche was its first Director, overseeing its expansion across England, Wales and Ireland as well as establishing a Geological Survey Museum, a School of Mines and a Mining Records Office. He was knighted in 1842 and remained Director until his death in 1855.

THE TWO WILLIAMS

In addition to Henry De la Beche, two other geologists with a close association with Lyme and Mary Anning were William Buckland and William Daniel Conybeare. Both were clergymen, both about ten years older than De la Beche and were established members of the Geological Society, Conybeare having joined in 1811, just four years after the Society was founded, and Buckland in 1813. Both had been educated at public school before attending Oxford University and becoming ordained into the Church of England. Despite De la Beche's antipathy towards organised religion, their shared interest in geology and strong Dorset connections allowed the three men to become friends and colleagues. Conybeare and Buckland were contemporaries at Oxford; De la Beche probably met Conybeare for the first time in Bristol in 1818; and he may have been first introduced to Buckland in Lyme, at the Assembly Rooms.[8]

WILLIAM BUCKLAND

William Buckland was born in Axminster, a carpet-making market town 5 miles (8 km) inland from Lyme over the Devon border. Like De la Beche, his

THE FOSSIL WOMAN

fossilising began when a boy, often accompanying his father, the Reverend Charles Buckland, on collecting expeditions along the coast. It is likely that on visits to Lyme in the 1790s, the young Buckland and his father would have encountered Richard Anning. Buckland regularly returned to Lyme whilst at Oxford. According to his son Frank, when fossiling on the beach 'crowds of little urchins ran after him to tempt him with pretty little golden serpents (pyritous ammonites) or wonderful thunderbolts (belemnites)'. His daughter, Elizabeth, later provided an account that gives a real sense of his personality.

> 'the vacations of his earlier Oxford time were often spent near Lyme Regis. For years afterwards local gossip preserved traditions of his adventures with that geological celebrity, Mary Ann Anning, in whose company he was to be seen wading up to his knees in search of fossils in the blue lias; of his breakfast-table at his lodgings there, loaded with beefsteaks and belemnites, tea and terebratula, muffins and madrepores, toast and trilobites, every table and chair as well as the floor occupied by fossils whole and fragmentary, large and small, with rocks, earths, clays, and heaps of books and papers, his breakfast hour being the only time that the collectors could be sure of finding him at home, to bring their contributions and receive their pay.'[9.]

He took his degree in 1805 and remained in Oxford, becoming a Fellow of Corpus Christi College and an ordained priest.

At Oxford Buckland attended the geology lectures of Dr John Kidd, Reader in Mineralogy, although these were not part of any formal university course – science degrees were not then available – and in 1813 succeeded Kidd as Reader in Mineralogy. Five years later he was appointed to the new position of Reader in Geology. Buckland soon developed a reputation as a lively, enthusiastic and popular lecturer employing maps, diagrams and specimens as well as humour to engage his audience, although some of his university colleagues thought Buckland's showmanship inappropriate for an Oxford don. Charles Darwin considered Buckland to be 'very good-humoured and good-natured,' but 'vulgar and almost coarse,' 'incited more by a craving for notoriety, which sometimes made him act like a buffoon, than by a love of science,' which is a rather unfair assessment. Buckland's excuse was his nervousness in front of a large audience, but once he had made them laugh he felt more relaxed.[10.]

After attending one of Buckland's lectures at the Geological Society, his former student Charles Lyell wrote to the fossil collector and Sussex doctor Gideon Mantell in 1822:

> 'Buckland, in his usual style, enlarged on the marvel with such a strange

mixture of the humorous and the serious, that we could none of us discern how far he believed himself what he said.'[11]

Buckland was undoubtedly the leading promoter and populariser of geology of the period and one of the greatest of all field geologists. The eccentricities for which he is best known diminish his achievements, especially his use of rigorous analysis to reconstruct past events. Stories abound of his antics: bringing a live bear in mortar board and gown to classes, one of a menagerie of wild animals he kept around the house; serving mice on toast to house guests; swallowing a French king's heart; and identifying saint's blood on the floor of a cathedral as bat's urine after tasting it. Many of these and other stories about him have no doubt been embellished and exaggerated in the telling, whilst Buckland himself seems to have enjoyed cultivating the manner of the eccentric professor.

In his inaugural lecture as Reader in Geology, Buckland set forth his belief that the facts of geology were reconcilable with the biblical record, and that there was ample geological evidence for 'a universal deluge'. Geology, he argued, was a science that supported the authority of the Bible and therefore an appropriate subject to be taught at Oxford. Buckland saw the hand of God in the design of the natural world and, like many of his contemporaries, he saw little contradiction between scripture and geological discovery. It was just a matter of how scripture was interpreted. Study of the natural world gave insights into God's plan; this applied not just to the study of geology and palaeontology where discoveries of new, strange and wonderful fossils were seen as wonders of God's creation, but across all the sciences.[12]

Buckland's geological researches were wide-ranging, initially directed towards developing an understanding of the Deluge or diluvial deposits, but also pioneering, at times taking an experimental approach to reach an understanding of the fossils he had discovered. On excavating broken fossil bones from a cave in Yorkshire in 1821 Buckland was able to prove, with the help of a hyaena from a travelling zoo and a side of beef, that the cave had been a hyaena den. On one occasion Buckland injected cement into the intestines of living fish to replicate some spirally-structured fossils. On another he woke his long-suffering (and understanding) wife in the middle of the night asking that she make pastry for the family's pet tortoise to walk over so that Buckland could better understand some fossil footprints. His published papers include a description of fossil bones from Oxfordshire belonging to a large, land-living reptile which he named *Megalosaurus*, the first published scientific description of what would later be classed as a dinosaur. He worked with De la Beche on the geology of Dorset and with Conybeare on the coal-bearing rocks of southwest England and South Wales.

Buckland's merger of scripture and geology is perhaps best illustrated by his 1836 book, *Geology and Mineralogy considered with reference to Natural Theology*. One of eight treatises funded by a bequest from the 8th Earl of Bridgewater to examine the 'Power, Wisdom, and Goodness of God, as manifested in the Creation', Buckland's 'Bridgewater Treatise' was the culmination of five years' work and summarised the state of geological and palaeontological knowledge as it stood in the early 1830s. Some of the information he included on the Lias fossils from Lyme came directly from the observations of Mary Anning.

In 1825, Buckland was appointed a Canon of Christ Church, Oxford, a position which gave him a comfortable sinecure, with a house, a thousand pounds a year and no university duties. His good fortune led Charles Lyell to remark with some envy to Gideon Mantell, 'Surely such places ought to be made also for lay geologists'.[13]

Now financially secure, Buckland married Mary Morland, a skilled naturalist and illustrator who was to collaborate with her husband on much of his research and writing. 'A most amusing, animated woman, full of strong sense and keen perception ... an admirable fossil geologist,' Mary Buckland also developed a friendship with Mary Anning and Elizabeth Philpot.[14]

Buckland remained in Oxford until 1845 when he was appointed Dean of Westminster, overseeing improvements to the Abbey and School, and lobbying for improved living conditions for the poor. He died in 1856, one of the best known scientists in Britain. As good an obituary as any is the *Elegy intended for Professor Buckland*, written in 1820 by Richard Whately, later Archbishop of Dublin:[15]

> Where shall we our great Professor inter,
> That in peace may rest his bones?
> If we hew him a rocky sepulchre,
> He'll rise and break the stones,
> And examine each stratum that lies around,
> For he's quite in his element underground.

WILLIAM DANIEL CONYBEARE

Although his family had property in Axminster, Willy (as his family called him) Conybeare was born in 1787 in London where his father was rector of a city church. His interest in fossils developed in childhood on family holidays to Bexley in Kent, now a London suburb. He went up to Oxford in 1805, meeting and befriending William Buckland at one of Dr John Kidd's geology lectures. Like Buckland, after graduating, he was ordained, first becoming a curate in Suffolk and then moving to St Luke's Church, Brislington, on the southern edge of Bristol in 1819. His father's death in 1815 left Conybeare wealthy, but he remained a clergyman, becoming vicar at Axminster in 1836 and in 1848 Dean of Llandaff Cathedral in Wales.[16.]

In contrast to Buckland's lively and outgoing persona, Conybeare was shy and reserved. Although the work he did thanks to Mary Anning is some of the most important in the history of palaeontology, they never forged the close relationship she enjoyed with Buckland and De la Beche, perhaps due his shyness. Tall and thin – gangly, even – his lecturing style was the antithesis of Buckland's. His 'ungraceful' manner was once described as frightening the ladies present, but he was undoubtedly Buckland's intellectual superior. When Dr John Kidd stood down from his position as Reader in Mineralogy in 1813, it was Conybeare who was approached as his successor. He declined, so the post was offered instead to Buckland. It was from Conybeare that Buckland learnt much of his geology, whilst the theological content of Buckland's inaugural lecture as Reader in Geology was largely Conybeare's. Conybeare advocated that a scholarly understanding of the Bible was necessary in order to maintain its relevance in the face of advancing scientific knowledge.

Conybeare and Buckland were members of a geological group, the Oxford Geology Club, and it was at a club meeting in 1818 at Clifton in Bristol that Conybeare met Henry De la Beche. The two were to form a fruitful partnership, working on fossil marine reptiles and in the establishment of the Bristol Philosophical Society and Bristol Institution in the early 1820s.

This period was a productive one for Conybeare; not only did he publish a ground-breaking series of papers on fossil marine reptiles, establishing his reputation as a palaeontologist, but in collaboration with geologist and publisher William Phillips he wrote an influential book describing the strata of England and Wales in 1822. *Outlines of the Geology of England and Wales* became a standard reference and the bible of many geological neophytes.[17.]

THE MISSES PHILPOT

The Annings were not the only fossil collecting family in Lyme; they shared their interest with three sisters who had moved to Lyme in about 1805 and had taken up fossil collecting as a hobby in about 1810, at about the time that Mary Anning sold her first ammonite. Margaret, Mary and Elizabeth Philpot lived together in a house bought for them by their brother John, a London solicitor. Like De la Beche, and despite differences of class and age, they befriended Mary Anning and may have helped guide the young Mary's fossiling interests. Of the three sisters, Elizabeth was the closest to

Mary, even though Elizabeth was almost twenty years older. The pair often went out onto the shore in search of fossils and worked together to identify what they had found. Although evidence is scanty, of all of Anning's friends Elizabeth Philpot was probably the most influential. In their letters, especially to Buckland and his wife, each often mentions the other, passing on messages and compliments and this is reciprocated in the replies. In a letter of 1 January 1834 to Elizabeth Philpot, for example, William Buckland asks her to find out from Anning what he should do with some specimens he had borrowed, while a letter from Mary Buckland of the same date expresses a hope that Anning had found 'something worth selling' in recent storms and asks Elizabeth to pass on her regards 'to Mrs Anning and Mary when you see them next time'.[18.]

The Philpot sisters built up a sizeable and scientifically significant fossil collection of about six hundred specimens which they kept in their house, Morley Cottage, at the top of Silver Street. The sisters and their collection were remembered by Lyme resident Salina Hallett in the 1920s:

> 'There has been a great deal written about Miss Anning, the fossilist. I can remember seeing her stand outside her shop in Broad Street. But I have never seen anything said about the two Miss Philpots that used to go fossilising with her ... They were great fossilists and had a very large collection. Several cases with glass tops and shallow drawers all down the front stood in the dining room and the back parlour and upstairs on the landing, all full of fossils with a little ticket on each of them. I heard they were worth £1000.'[19.]

Salina Hallett also recalled that

> 'They were very good kind ladies to poor people. One thing they were noted for was their home-made salve, of which they always kept a large store for anyone that chose to go and ask for it, and they had a great many applications for it, as it was so good for any sort of wounds.'

Woodville, the house on Silver Street leased by De la Beche's stepfather is close to Morley Cottage, so from 1814 De la Beche and the Philpots were neighbours. De la Beche often accompanied Elizabeth Philpot and Mary Anning on their frequent fossil-hunting trips to the beach.[20.]

The Philpots were generous in allowing access to their collection, welcoming visitors who wished to examine it, and lending specimens for research and publication. Some of the most attractive fossils from the Lias of Lyme and Charmouth are the coiled, spiral shells of ammonites, extinct molluscs related to squid and octopus – of which the Philpots had some fine specimens. Often beautifully preserved and prepared, ammonite fossils sell well today often taking pride of place in the fossil dealers' shop windows in Lyme; it would have been the same in Mary Anning's day, especially as many were new species. The sale of small fossils like ammonites to Lyme's visitors would have provided much of Anning's income, supplemented occasionally by the sale of a large marine reptile skeleton.

Some of the best-known ammonites from Dorset were first described by James Sowerby who began issuing a part-work on fossils in 1812, *The Mineral Conchology of Great Britain*. After Sowerby's death in 1822, publication was continued by his son, James de Carle Sowerby until 1846. Beautifully illustrated with hand-coloured plates, *The Mineral Conchology* is still used as a reference work today by palaeontologists. The fossils illustrated were given or loaned to the Sowerbys by collectors such as William Buckland, De la Beche and Elizabeth Philpot. She lent James Sowerby a specimen which he illustrated in 1817 and named *Ammonites obtusus*. This fossil is now called *Asteroceras obtusum* and beautiful examples can still be found. Sowerby was tardy in returning the specimen and Elizabeth Philpot had to ask Buckland to pick it up from Sowerby the next time he was in London, which he forgot to do. In the end, Sowerby sent it to Elizabeth's brother, John, at his solicitors office for him to take to Lyme. This sort of arrangement was typical of the connections within the developing network of geologists and collectors at that time.[21]

MISS CONGREVE AND MRS KENNAWAY

Also living in Lyme and collecting fossils was a 'Miss Congreve' who provided ichthyosaur specimens from her collection to De la Beche and Conybeare. This was probably Miss Ann Congreve who we know was a subscriber to J.S. Miller's 1821 book on fossil crinoids and who rented Sherborne House in Broad Street from 1806. Her sister, Mary, may also have had an interest in fossils. The sisters died in 1823, but they would certainly have known Mary Anning as a young woman, as well as the Philpots.[22]

A little more, but not much more, is known of Sarah Kennaway, a 'lady of means' of Charmouth. Born Sarah Johnson in Middlesex in 1767, she was the wife of Robert Kennaway, an Exeter merchant. The couple moved to Charmouth some time before 1803 and by 1822 Sarah had developed an interest in geology and become friends with Mary Anning. In response to a request from Sarah about fossils found around Lyme, Anning sent her a long list of those that she could remember. Her letter also acknowledges a gift Mrs Kennaway had sent Anning and discusses mutual health concerns, an indication of the closeness of their friendship. It would seem that Mary Anning was unwell – the nature of her illness is unclear – and that Sarah was arranging for her to see a doctor in Exeter, as Anning writes:

> 'I am very sorry to hear your journey to Exeter, has been of so little service to you, I fear you will not be better till the spring, as I cannot think this unseasonable weather is good for invalids, I am greatly obliged to you for your kind present, and also for your kind intentions of speaking to your doctor but I was too ill to undertake the journey, it seems an age since I had the pleasure of seeing you.'[23]

Elizabeth Philpot mentions Mrs Kennaway in an 1835 letter to William Buckland, so she was also known to them. The letter hints at mustering financial help for Mary Anning, so their friendship continued into the mid-1830s, and probably longer.

ENTER COLONEL BIRCH

One of the Annings' clients was a retired officer in the Life Guards, Lieutenant-Colonel Thomas James Birch. Born in London in about 1768, like many fellow officers, he may have retired or been placed on half pay after the Battle of Waterloo in 1815 brought peace. By then he had taken up fossil collecting, mainly through purchase, and toured the West Country, including visits to Charmouth and Lyme, buying specimens. Three years after Waterloo he acquired, probably from the Annings, the most complete ichthyosaur skeleton yet found, and the first to show all four limbs. For the first time the shape of the body and the four paddles were apparent. De la Beche had the specimen sent to Everard Home at the Royal College of Surgeons for examination. In a letter acknowledging its arrival, Home told De la Beche that

> 'It is everything that could be desired, as it explains all the parts of the skeleton not already made out. It is highly valuable as being unique, and certainly should have a place either in the British or Hunterian Museum' and that it is 'one of the most extraordinary animals that inhabited the Antediluvian world.'[24]

Birch's collection included various ichthyosaur specimens, fossil fish, crinoids, and ammonites, many of them purchased from the Annings. One of his ammonites from Lyme was described by the Sowerbys in 1820 and named *Ammonites birchi*. Today it is called *Microderoceras birchi* and gives its name to part of the rock sequence on Black Ven.

In the summer of 1819, Birch again visited Charmouth and Lyme, only to find the Annings about to sell their furniture in order to pay their rent. Despite the income from their first ichthyosaur, the Annings had still needed help from parish poor relief until at least 1816. They had made some further sales, probably to James Johnson, Birch and De la Beche amongst others, but they had discovered little worth selling since the middle of 1818, so by the end of the decade the Annings were again struggling financially. Whenever in Lyme, Birch spent time with them, not only in the fossil shop but also out on the beach fossiling with Mary, and had grown fond of the family. On seeing their predicament, he resolved to try and help them.

FIVE

The First Plesiosaurs

IN JULY 1819, while fossil collecting in Sussex, Colonel Birch called on the doctor and fossil collector, Gideon Mantell, in Lewes to view his collection. Although the two had never met, Mantell thought Birch 'a very agreeable and intelligent man'. The colonel was on his way to Dorset, and Mantell asked if he might be able to help him acquire some ichthyosaur specimens from Lyme for his collection. On 5 March 1820 Birch wrote to him to say:

> 'I have not forgotten my promise to select for you some fine things from the blue lias – I cannot however, perform it yet as I have great occasion for every individual specimen I can muster. The fact is that I am going to sell my collection for the benefit of the poor woman and her son and daughter at Lyme who have in truth found almost <u>all</u> the fine things, which have been submitted to scientific investigation: when I went to Charmouth and Lyme last summer I found these people in considerable difficulty – on the act of selling their furniture to pay their rent – in consequence of their not having found one good fossil for near a twelvemonth. I may never again possess what I am about to part with; yet in doing it I shall have the satisfaction of knowing that the money will be well applied, the sale is to be at Bullock's in Piccadilly the middle of April. Should you then be in town don't miss seeing it.'

This was a remarkable and generous gesture by Birch. It was his intention to sell his entire collection of Lias fossils, many probably purchased from the Annings in the first place, to help ease their poverty. Mantell did as Birch suggested and attended the sale, noting, much later, that 'it was subsequently understood that all the most valuable fossils had been obtained by the indefatigable labours of Miss Mary Anning.' James Sowerby in *The Mineral Conchology* commented that Birch's 'generous method of disposing of his collection will long be remembered'.[1]

The sale took place on 15 May 1820 at the Egyptian Hall in Piccadilly, where only a year earlier the Annings' first ichthyosaur had been sold when Bullock's museum collection was auctioned. Birch's sale, with Bullock as auctioneer, comprised 102 lots. The sale included, as Lot 60, an ichthyosaur skull which Home had used in one of his papers: this sold for 14 guineas to the Musée National d'Histoire Naturelle in Paris which also purchased a number of other ichthyosaur specimens. Another skull fetched £11 15s, and a partial skeleton £17 6s 6d. The sale realised over £400 and although we don't know exactly how much Birch gave to the Annings, there is no reason to suppose that it was significantly less than this.[2]

The sale's star specimen, Lot 102, failed to sell. This was the complete ichthyosaur illustrated by Home in 1819 and brought to his attention by De la Beche. Birch, it seems, wanted £300 for it and when the bids failed to reach his reserve, it was bought in. It was subsequently sold to the Royal College of Surgeons for £100, so Home's wishes, expressed in his September 1818 letter to De la Beche, that the specimen should be placed in the Royal College of Surgeons' Hunterian Museum were eventually realised.

Birch's generosity raised a few eyebrows, most notably those of the Bristol fossil collector, George Cumberland who wrote on 20 August 1820, 'Mrs Hanning is the dealer at Lyme. Col. Birch is generally at Charmouth (they say Miss Anning <u>attends</u> him),' implying that there was more to the relationship between Mary Anning, then aged 21, and Birch at 52, than simply vendor and buyer. There is no other evidence to support this insinuation.[3]

Both Birch's and Cumberland's references to the Annings imply that Molly was running the fossil business. Theirs was a family business, with all three members involved. Mary was certainly helping with sales by 1820, for that summer the Reverend Adam Sedgwick, Woodwardian Professor of Geology at Cambridge University, paid the Annings a visit.

Sedgwick is yet another of the group of scientists whose collections were enhanced by Mary Anning, and who became one of her main customers. Born in the Yorkshire Dales in 1785, Sedgwick went to Trinity College, Cambridge and in 1818 was both ordained as priest and appointed to his professorial post at Cambridge, despite having only a limited knowledge of geology, although he had attended meetings of the Geological Society as a visitor since 1816. He served as the Geological Society's President from 1829 to 1831, and remained Woodwardian Professor until his death in 1873.[4]

Sedgwick's researches were focussed on the older rocks of the Lake District and North Wales. In 1831 he was accompanied in his North Wales fieldwork by a young Charles Darwin, teaching him geological mapping. Sedgwick's work in Wales led to the recognition, in 1835, of one of the periods of geological time, the Cambrian Period. A year later, with geologist

Roderick Murchison, he defined another period of the geological column, the Devonian. Despite his friendship with Darwin, he opposed Darwin's ideas of evolution through natural selection, adhering firmly to his conviction that a series of divine creations had given rise to different species.[5]

In 1820, just two years into his geological career, Sedgwick was still learning his trade and had set out on a geological tour of the West Country. After exploring the Mendip Hills with William Daniel Conybeare, he continued down to the Devon coast at Sidmouth with a view to working his way along the south coast to Portsmouth before returning to Cambridge. On 20 September 1820, he noted in his field journal: 'After breakfast purchase fossils of Miss Anning'. This he did, spending £3 2s on 'various fossils' and a 'part of Ichthyosaurus', according to a receipt signed by Mary Anning. That it is 'Miss Anning' and not 'Mrs Anning' that Sedgwick notes he dealt with, suggests that it was now twenty-one year old Mary who is dominant in the business, and this is supported by contemporary newspaper reports mentioning 'Miss Mary Anning' by name.[6]

Her discoveries meant that by now she had a good working knowledge of the local geology. In 1822 she provided her friend Sarah Kennaway in Charmouth with a comprehensive list of all of the fossils and minerals that she could recall being found in the various strata along the coast between Charmouth and Axminster as well as in the more recent river gravels which she knew contained mammoth tusks, the teeth of rhinoceros and 'a species of Bullock'. Although she lists ammonites, *Nautilus* and belemnites, she glosses over other shelly fossils and admits she 'dont know any thing of shells'. Her listing of 'Blue Lias, Ichthyosarus Vulgaris Ichthyosarus platydon Tenuirostris, Plesiosarus, dorsal fins resembling the radie of Balistis, dapedium politum, fragments of three other distinct species of fish' demonstrates that she was more confident in her knowledge of fossil vertebrates.[7]

Joseph, although employed as an upholsterer, was also still involved. His signature, with the date of 12 May 1820, is on a copy of the Birch sale catalogue with a note that 'This Catalogue belongs to the Fossil Shop Lyme' – proof that by 1820, the Annings had set up business as 'The Fossil Shop' at their home in Cockmoile Square. Later correspondence suggests that Joseph remained involved until at least 1832.[8]

The start of the 1820s saw Henry De la Beche begin to pull together his research on ichthyosaurs in cooperation with William Conybeare. Amongst the fossil collectors in and around Bristol, was a Brislington neighbour of Conybeare's, George Weare Braikenridge. Braikenridge's collection included an ichthyosaur skull, described as a crocodile, found in 1813 in a quarry near Keynsham in Somerset. Also in Bristol, at Hotwells, was James Johnson and his collection of specimens from Charmouth and Lyme, which was 'very extensive, though not at all arranged, and deficient in many things'; George Cumberland and Richard Bright; and William Morgan at Bower Ashton. At nearby Bath was Charles Hunnings Wilkinson, and at Hinton Charterhouse, Samuel Skurray Day.[9.]

De la Beche had examined some of these collections in the spring of 1819 while working on the different species of *Ichthyosaurus* he recognised from Lyme. He and Conybeare exchanged letters as they worked out the skeletal anatomy of these reptiles and produced a paper which Conybeare presented at a meeting of the Geological Society in April 1821. While De la Beche sought out material for study, Conybeare methodically pieced together the specimens, disentangling the bones and making sense of the structure of the skeleton. Based on specimens, including twenty ichthyosaurs with skulls, from Lias rocks in Somerset, Gloucestershire and Leicestershire as well as material found in Dorset, they described three species of *Ichthyosaurus* and added much new detail on the anatomy of these animals. In their paper they dismissed Home's name of *Proteosaurus*, saying that the fossil's features were too dissimilar to the salamander to warrant the name, so they retained the name *Ichthyosaurus*. None of this went down well with Sir Everard Home. Buckland reported to De la Beche in November 1821 that he had 'seen Sir E.H. who is out of humour about your Paper & naturally enough'. This was no surprise; in a postscript to a letter to Buckland in March 1821, Conybeare had told him. 'I expect Sir E.H. will be as mad as can be'. Home, the distinguished Professor of Anatomy and Surgery at the Royal College of Surgeons, had laboured for five years trying to make sense of these fossil bones – work now being cast aside by two upstart geologists half his age.[10.]

Amongst the fossil material examined by Conybeare and De la Beche they had identified fossil bones which were not ichthyosaur, but represented another new, related animal. The vertebrae were longer than those of ichthyosaurs and did not show the same hour-glass cross-section, resembling crocodile vertebrae more than ichthyosaur. The small bones of the paddle-shaped limb were also completely different from those of ichthyosaurs.

This new discovery became the main theme of their paper, 'Notice of the discovery of a new fossil animal, forming a link between the Ichthyosaurus and crocodile', even though they were dealing with only incomplete and fragmentary specimens. De la Beche had a specimen of eighteen linked vertebrae from the middle of the animal's back while Colonel Birch had sixty-three vertebrae, but these had been collected mostly as loose bones and some of them were missing. He did, however, have part of one of the front limbs. It is almost certain that Birch bought his specimen from the Annings, although we have no record of where and when it was discovered.[11]

Conybeare and De la Beche found other specimens in the collection of the Bristol Library Society, collected by a Bristol vicar, the Reverend Alexander Catcott in the mid eighteenth century – more vertebrae plus a shoulder blade and limb bone. They recognised that a specimen described and illustrated by the antiquarian William Stukeley in 1719 also belonged to their new animal, as did material found at Whitby in Yorkshire. None of the specimens, though, preserved the skull. They were on the brink of making one of the most important discoveries in the emerging science of palaeontology.[12]

In the manuscript of their joint paper, Conybeare named the new animal *Engistosaurus*, from the Greek *engistos*, 'nearest' and *sauros*, 'lizard', to recognise its closeness to crocodiles. As the best specimens were in Colonel Birch's collection and as he considered all the material to be just one species, Conybeare named it *Engistosaurus Birchii*. Shortly before before publication, he had second thoughts, crossing out *Engistosaurus* and substituting *Plesiosaurus*, from *plesion*, 'near', and dropped the species name from the final published paper.[13]

Over the course of the next year, De la Beche and Mary Anning searched for a complete plesiosaur, to no avail. Further, incomplete, material was discovered in Somerset: a skull at Street by Thomas Clarke and a jaw from near Bath by Colonel Birch, while De la Beche also found part of a jaw. In a follow-up paper in May 1822, Conybeare pointed out that they now had, from the same rocks, skeletons without skulls, and skulls without skeletons: 'We find in the same place skeletons of a Saurian animal wanting the jaw, and the jaw of a Saurian animal wanting the other bones; and no other claimants exist for either.' Since there were no other candidates, Conybeare proceeded on the basis that one belonged to the other. These two detailed papers by De la Beche and Conybeare put the study of these marine reptiles on a scientific footing.[14]

With much scientific attention focussed on these remarkable fossils, the fame of the Annings, and in particular that of Mary, spread, with monthly news magazines reporting her ichthyosaur discoveries and mentioning her by name. In 1821 it was reported that

> 'a complete specimen found ... a few days since, by Miss Mary Anning, of Lyme ... Its length is five feet ... another fossil of a similar description was found by Miss Anning about six weeks ago near the same spot. The remains of this beast measure nearly 20 feet in length; its vertebrae are 95 in number; its head five feet in length.'[15]

Some, however, either failed to recognise her gender or confused her with her brother, assuming that such wonderful discoveries could only be the work of a man:

> 'Mr. Anning, of Lyme, who has made so many interesting discoveries in the Fossil World in that neighbourhood, has now in his possession one of the lizard tribe that is not to be found in any European cabinet. It was discovered imbedded in blue lias, near Lyme, and is now in Mr Anning's well-furnished Fossil Repository, in that town. It is in so perfect a state that its osteology is clearly developed.'[16]

Others read almost like an advertisement, and one senses the hand of George Roberts:

> 'There are now at Anning's Fossil Depôt, Lyme Regis, three fossil skeletons of the Saurin tribe, viz. Ichthyosaurus Tenuerostris, Ichthyosaurus Vulgarus, and Ichthyosaurus Intermedius; the former being twelve feet in length, and in such a perfect state that its osteology may be distinctly ascertained. The Ichthyosaurus Vulgarus is a beautiful cabinet specimen, unequalled by any hitherto found in Europe, being only three feet long.'[17]

> 'Miss Mary Anning has collected a treat for the fossilist, and admirers of the wonders of an antediluvian world, in three beautiful specimens lately found: one an Icthyosaurus Intermedius, two Ichthyosauri Vulgares ... Miss Anning has found a Cornua Ammonis of unusual size and beauty.'[18]

In May 1821, the Annings found perhaps the best preserved ichthyosaur yet discovered, a 5 feet (1.5 m) long skeleton for which they asked £100. De la Beche sought to have it purchased by the British Museum and in a letter of 11 July 1821, told Charles Konig at the Museum that

> 'the Annings who search for fossils here had found a very beautiful small skeleton of *Ichthyosaurus communis* exceeding in preservation any yet found. I immediately obtained the refusal of it for the British Museum.'[19]

But refuse it they did, opting instead for a less well-preserved example found a year or two earlier at half the price. Even this they considered expensive, delaying payment until prompted by Molly Anning in a letter to Charles Konig on 2 September 1821:

> 'Sir, I hope you will not be offended by my addressing you on the subject of a Fossil which I had the honour of sending up to the British Museum at the desire of Mr Buckland for the sum of fifty pounds, which I hope you have received safe, I am very sorry to hear that the fossil is considered dear, the same sum was offered for it before Mr Buckland saw it I shall be very happy

to make a difference in the price of any other Fossil that I may find hereafter and which may be thought good enough to be purchased by the British Museum, As I am a widow woman and my chief dependence for supporting my family being by the sale of Fossils I hope you will not be offended by my wishing to receive the money for the last Fossil as I asure you Sir I stand much in need of it.'[20.]

Such is her urgency that Molly Anning is offering Konig a discount on a future purchase in compensation for their current purchase being considered too expensive. The letter suggests that she was running the business, although she may be exaggerating to gain his sympathy; *she* sent the fossil to the Museum; *she* is supporting her family; and is offering a discount on a fossil that *she* might find in future.

The specimen which the Museum declined, later described as 'the *very finest specimen of a Fossil Ichthyosaurus ever found in Europe*' was subsequently purchased from the Annings for £50 and presented on 4 January 1823 to the new Bristol Institution for the Advancement of Science and Art, by a group of nine donors which included Conybeare and De la Beche.[21.]

Another letter demonstrates Joseph Anning's continued involvement. In 1825 he wrote describing additional ichthyosaur material being offered for sale to the Museum:

'Sir Some time since my Sister wrote you, concerning the Ichthyosarus Tenuirostris, which you thought proper to refuse, for reasons which you assigned Sir, I write now to offer to your notice, a Specimen of the Ichthyosaurus which I have in my possession, about 3 foot in length, very perfect. Head & eye beatifully preserved, with the last Vertebrae of the Tail, paddles & ribs, equally fine in fact it so perfect that its Osteology can be clearly ascertained a very fine Cabinet specimen, its weight not more that 30 or 40 pounds price 20 Guines. Answer for return will oblige'.

Konig may have changed his mind and taken the *Ichthyosaurus tenuirostris* as there is a specimen, now named *Leptonectes tenuirostris*, which is thought to have come from Anning in the Natural History Museum.[22.]

But as a woman, and still young, it was on Mary that most attention was focused. George Cumberland, writing in the *Bristol Mirror* in January 1823 about the Bristol Institution's new specimen said that its discovery was due to 'the persevering industry of a young female fossilist, of the name of Hanning of Lyme in Dorsetshire, and her dangerous employment' and gives us a vivid picture, albeit with some exaggeration, of the hazards she faced:

'This persevering female has for years gone daily in search of fossil remains of importance at every tide, for many miles under the hanging cliffs at Lyme, whose fallen masses are her immediate object, as they alone contain these valuable relics of a former world, which must be snatched at the moment of their fall, at the continual risk of being crushed by the half suspended fragments they leave behind, or be left to be destroyed by the returning tide: – to her exertions we owe nearly all the fine specimens of Ichthyosauri of the great collections; and, to shew that it is one which rewards industry a single specimen of her's, *far inferior to this placed in the Institution* was lately sold to the College of Surgeons for the sum of One Hundred Pounds.'[23]

Mary Anning's principal champion, however, was George Roberts. In 1823, Roberts published his comprehensive *History of Lyme Regis* which also included chapters on Lyme's climate and natural history as well as a *Geological notice of the coast and a description of the fossils* based largely on the published work of De la Beche, Conybeare and Buckland. But Roberts is too good a storyteller to omit the lightning strike of 1800 and the transformation of the dull child to an intelligent one who 'as a fossil-hunter, was destined to bring to light some of the grandest relics of a primeval world that have been discovered in any age or country'. He doesn't mention Mary Anning by name, although it would have been clear to many of his readers who he was writing about. The fossil shop too, receives only a passing reference. Where she does get a mention is amongst the more than two hundred subscribers to Roberts' book where she is listed as 'Miss Mary Anning, Lyme'. Books were expensive in the 1820s, though Roberts' volume sold for a modest seven shillings, this still would have been a significant outlay for Anning. But there she is, listed amongst the local gentry and aristocracy, including some of her friends and clients such as De la Beche, Miss Philpot (probably Elizabeth), Dr Carpenter, and James Johnson of Bristol. Her copy, which she signed and dated on the title page, survives in Cardiff University Library.[24]

ANNING'S EAGLE EYE

In her search for fossils Anning took advantage of the work of the natural agents of weathering and erosion. Winter storms brought rough seas and large waves which undercut the Lias cliffs and cause their collapse. Heavy rain could lubricate the soft mudslides of Black Ven and start them moving downslope. Even twice-daily high tides could expose new fossils through erosion and scouring, removing sand and mud from the beach to expose fossils in the bedrock or to wash them out of the mudflows. Setting out from the warmth of Cockmoile Square as the tide began to ebb, Anning would begin her search. The smaller shelly fossils like belemnites and ammonites were relatively easy to find and collect, perhaps weathered out, or needing simply a well-placed tap from a hammer to break open a block of rock, the edges of which may have displayed only a hint of its fossil content. Anning developed an expert eye for this, and recognising a fossil-bearing rock from the subtlest of clues. The writer and lecturer John Murray once went fossilising with her:

> 'I once gladly availed myself of a geological excursion and was not a little surprised at her geological tact and acumen. A single glance at the edge of a fossil peeping from the Blue Lias revealed to her the nature of the fossil and its name and character were instantly announced.'[25.]

Another visitor to Lyme also commented on Anning's ability to spot a fossil:

> 'She has more than the power of an eagle's eye when she searches among the sands and rocks, and can distinguish when fossils are enclosed in the stone.'[26.]

A trace of fossil bone in a rock might indicate merely a single vertebra or fragment or, more rarely, the start of something big: a partial or complete skeleton of a marine reptile. These larger specimens were more of a challenge to collect. The individual bones were not extracted from the rock at the discovery site, but lifted still embedded in perhaps half a dozen or more, possibly large, heavy blocks for later, more leisurely removal back in the workshop. Loosening these blocks from the bedrock was slow patient work with hammers, chisels and levers. Depending on where on the foreshore the specimen was found, it was often impossible to complete the recovery work before the tide turned, flooding in over the rocks and covering the partially exposed specimen. All that Anning could do was wait, and hope that it would still be there when the tide went out again. Once the rock blocks were detached, the next challenge was to transport them back, perhaps several miles along the beach, to her workshop at home. For this she almost certainly needed help.

In the early nineteenth century, the foreshore rock ledges and cliffs at Lyme that were the source of Anning's specimens were being actively quarried for building stone and for the manufacture of stucco. The mineral rights for most of this section of the coast, including Church Cliffs, were owned by the same Henry Hoste Henley who purchased the 1811-12 ichthyosaur from the Annings. The natural erosion of the coast and the quarrying laid bare a wide extent of fresh limestone beds which Anning could examine. Discussing ichthyosaurs, George Roberts in his 1839 dictionary of geology noted that

> 'The great depository is Lyme Regis: the reason is, that a greater extent of

Lias is there acted upon by the tide, and men, who break up the ledges; and so enable Miss Anning to perambulate a fruitful superficial extent of three miles long by one eighth of a mile broad.'[27]

The quarrymen themselves would have come across fossils, large and small, during the course of their work manually breaking open the beds of limestone. No doubt they ran a lucrative sideline in selling these on to Anning for sale in her shop. She would also have benefitted from the quarrymen's news of their discoveries and from their stone-breaking skills and strength, seeking their assistance to extract and move the larger fossils.

The quarrymen moved their rock from the beach on the backs of donkeys or by using stoneboats, heavily-built wooden rowing boats capable of carrying six or seven tons. Once loaded these were then rowed by two men to the Cobb where the stone could be stockpiled or transferred to a ship for transport to London. This may be how some of Anning's larger reptile specimens were moved, the blocks being unloaded at the Cobb or onto the beach in front of The Walk, now Marine Parade, and then taken to Cockmoile Square by cart or donkey.[28]

Not all of Mary Anning's discoveries were so convenient as to be found on the beach. Some came from high on the cliff. Sections of Church Cliff tower up a sheer 200 feet or so, making access all but impossible. Despite its name, the upper part of Black Ven is not so forbidding. Here the softer mudstones and frequent landslides form a sloping cliff which can be climbed, although only with great care: it is all too easy to slip or become trapped in thick mud.

If the landward side was fraught with hazards, so also was the seaward. Anna Maria Pinney who went out collecting in the early 1830s with Anning on the beach at Church Cliff where she found many of her best specimens gives a vivid account of its perils:

> 'She was out just before the waters begin to ebb, and we climbed down places, which I sd [should] have thought impossible to have descended had I been alone. The wind was high the ground slippery and the waves beating against the Church Cliffs as we went down. Our dangers were by no means over, for when we had clambered to the bottom of the Corporation wall, we had frequently to walk along the blue Lias cliffs, where there was just room to stand and no more the sea being behind us. In one place we had to make haste to pass between the dashing of two waves, before I knew what she meant to do, she caught me with one arm round the waist, and carried me for some distance, with the same ease as you would a baby.'[29]

This last remark suggests that years of lifting, moving and carrying heavy fossil specimens had given Anning a particular physical strength.

Once a specimen was back in her workshop, Anning could begin the slow and painstaking task of exposing the fossil from the rock in which it was embedded. Again using hammers and chisels, but lighter in weight than those needed on the beach, she would chip or delicately scrape away – she called it 'picking' – at the rock matrix to reveal as much of the fossil as she dare. Remove too little and the fossil might not be suitable for scientific study or for attractive display; too much and it might become too fragile to move and easily broken. This skill she would have learned as a girl from her father, and refined it with years of practice. It needed a clear understanding of the anatomy of these fossil animals to know what was fossil and what was rock, which bits to retain and which to remove, what was the fossil's shape and where did it lie in the rock. In this she was expert. George Roberts praised her 'great judgement in extracting the animals, and infinite skill and manipulation' when removing the rock matrix from the bones. Working in the dim light and dust of her workshop was tiring on her eyes, leaving them at times too inflamed for her to read, write or draw.[30.]

The complex skeletons of the marine reptile fossils presented a particular jigsaw-like challenge, in that she had to fit the blocks of bone-bearing rock together into the positions in which they were found, to show the whole fossil as it lay in the ground. Often this was not in the exact position the individual bones had occupied in life. Some might have been displaced and scattered after death, but before burial and fossilisation. She was not reconstructing or restoring the skeleton; she recognised the importance to science of displaying the fossil as it was preserved in the rock. In order to do this, she carefully positioned the bone-bearing blocks within a wooden frame constructed, no doubt, with the aid of her cabinet-maker brother, and then set the blocks in place using plaster of Paris. The end result would resemble a picture in a frame which could be mounted on a wall. One disadvantage of this method is that only one side of the specimen is visible; another is that the whole thing can be extremely heavy and difficult to move.

PLESIOSAURUS

Since Conybeare and De la Beche first identified the presence of another fossil reptile, *Plesiosaurus,* in the Lias rocks in 1821, the search had continued for a complete specimen. Success finally came on Wednesday 10 December 1823 when fossil bones were discovered by Mary Anning on the beach below Black Ven. The newspapers carried the story:

'Mary Anning, the well-known fossilist, whose labours have enriched the British Museum, as well as the private collections of many geologists, lately found, east of the town, and immediately under the celebrated Black Ven

1. Lyme Regis from the northeast in about 1850, with St Michael's Church (left) and the Cobb (centre left) above the old lower town in the valley of the River Lim with its cloth mills.

2. Lyme from the Cobb, 1840s. Some of Mary Anning's larger specimens were sent to London by sea from here.

3. Cobb Gate from Bell Cliff before the fire of 1844 which destroyed many of the buildings between the Three Cups Hotel (on left with upper bay windows) and St Michael's Church, in the background. To the right is the roof of the Assembly Rooms. In the centre is narrow Bridge Street and the Buddle Bridge where Mary Anning was almost crushed by a cart in 1833.

4. By the late nineteenth century the building on the corner of Bridge Street at Cobb Gate had become the Fossil Depot. A well-known Lyme landmark, and much photographed, it had no connection with Mary Anning.

5. Lyme, looking west from near Church Cliffs, 1825. Gun Cliff is left of centre, the buildings above opening onto Cockmoile Square which is hidden in this view. The building far left is the Assembly Rooms.

6. The Anning's house in Cockmoile Square which they occupied from about 1808 to 1826. This is the only known illustration of the house and was drawn in 1842 by W.H. Prideaux and E. Liddon, students at George Roberts' school.

7. Lyme Regis Museum in Cockmoile Square today. The wing with the two arches probably occupies the site of the Anning's house.

8. A view to the Cobb, an 1825 lithograph by Charmouth artist Thomas Carter Galpin. The bonneted woman in the foreground gazing out to sea and holding a hammer and a fossil may, or may not, be Mary Anning.

9. The Walk (now Marine Parade) and Cobb Hamlet from near the Assembly Rooms, Lyme Regis, about 1833.

10. A view of the Cobb and harbour from Langmoor Gardens.

11. A view down Broad Street, about 1900. Mary Anning's shop was the low two-storey building centre right, the second down from Dunster's shop. The dark shop front with its oval sign probably dates from the 1860s, but the upper storey with its bay window may have been unchanged since Anning lived there.

12. The Congregational Chapel, Coombe Street, where Mary Anning was baptised and educated.

13. Plaque marking the site of Mary Anning's fossil shop on Broad Street.

14. Cobb Gate, site of the Assembly Rooms, with St Michael's Church (upper right) with the roof cupolas of the Guildhall and Lyme Regis Museum in Cockmoile Square below.

15. Mariners Hotel, Silver Street, Lyme which includes Morley Cottage, home of the Philpot sisters, and 16. Woodville (left), home of Henry De la Beche between 1814 and 1822, Silver Street, Lyme Regis.

17. Lyme Regis from the Cobb. The beach huts in the background mark the line of Marine Parade, and the sweep of coast to the right of the town extends to just beyond Charmouth.

18. Geological map of the Devon and Dorset coast between Beer (left) and Charmouth (right) by Henry De la Beche. Lias rocks are shown in purple with the overlying beds of Greensand shown in green. From George Roberts, 1834, *The History and Antiquities of the Borough of Lyme Regis and Charmouth.*

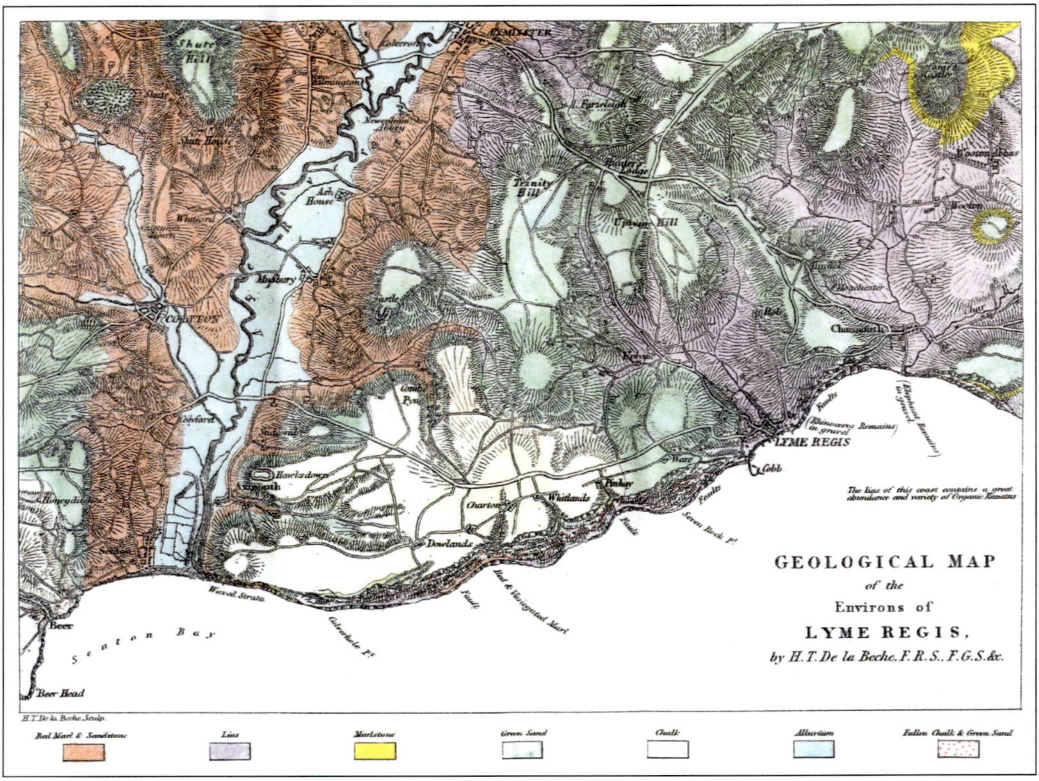

19. Coast east of Lyme: Black Ven, Charmouth (in valley), Stonebarrow Hill and Golden Cap.

20. Blue Lias Formation limestones and mudstones, Lyme Regis.

21. Lias limestones and mudstones on the landslipped slopes of Black Ven, with Church Cliffs (left) and Broad Ledge (foreground). Cretaceous Greensand caps the top of the hill.

22. Large ammonite in Blue Lias Formation limestone, west of Lyme Regis.

23. Mary Anning, oil painting probably by William Gray, February 1842, a few months before her 43rd birthday, with her dog Tray.

24. Mary Anning's sketch of 'my old faithful dog', a predecessor to Tray, the victim of a rockfall in 1833.

25. Mary Anning, pastel drawing by B.J.M. Donne, 1850, a copy of Gray's painting.

26. Henry De la Beche (1796-1855), aged about 22 and when he was living in Lyme.

27. William Daniel Conybeare (1787-1857) in 1824, with a drawing of the complete plesiosaur discovered by Mary Anning.

28. Sir Everard Home (1756-1832), comparative anatomist at the Royal College of Surgeon's museum.

29. Adam Sedgwick (1785-1873), Woodwardian Professor of Geology, Cambridge University in 1833.

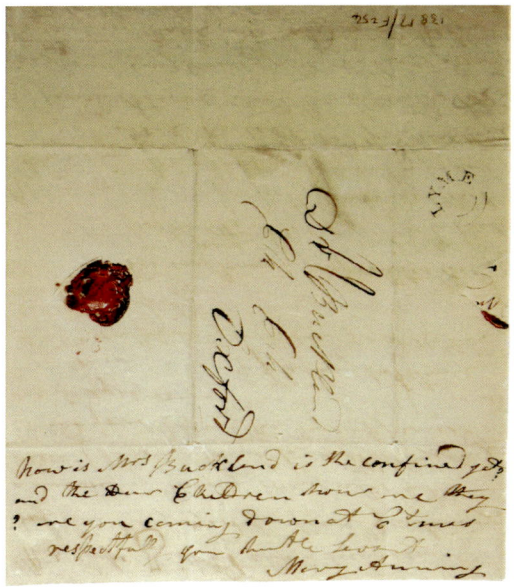

30. Part of a letter of 21 December 1830 from Mary Anning to William Buckland. Anning wrote with a strong, confident hand.

31. William Buckland (1784-1856), in 1823. The ichthyosaur skull in the foreground belonged to Henry De la Beche.

32. William Buckland, holding an ammonite, lecturing in 1823 in what is now the Old Ashmolean Museum in Oxford.

33. Richard Owen (1804-1892), palaeontologist and comparative anatomist at the Royal College of Surgeons, in 1840.

34. Roderick Impey Murchison (1792-1871) in 1840. Mary Anning thought him 'certainly the handsomest piece of flesh and blood I ever saw'.

35. Introduction and first verse of a poem about Roderick Murchison hand-copied onto decorative notepaper by Mary Anning from *Bentley's Miscellany* late in 1846 or early 1847.

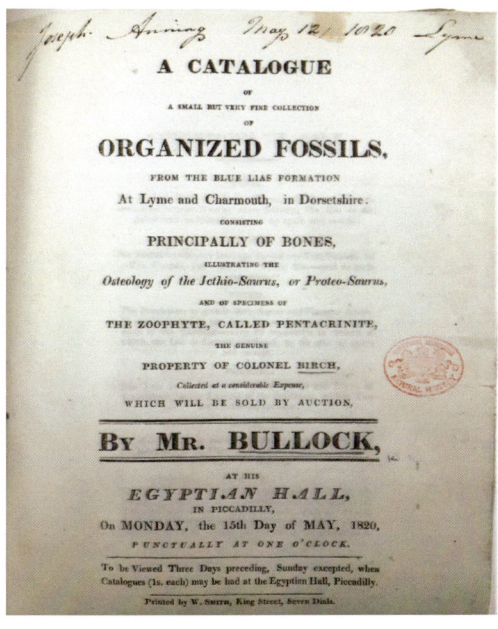

36. William Bullock's London Museum in Piccadilly where the first Anning ichthyosaur was displayed.

37. The Anning fossil shop's copy of the 1820 auction catalogue for the the sale of Colonel Birch's collection, signed by Joseph.

38. Cabinet-maker's try square, inscribed 'Anning' which may have belonged to Joseph Anning (left); trowel said to have belonged to Mary Anning (centre); 'Mary Anning's fossil extractor', actually the handle of an 1882 British Army spade (right).

39. Perhaps the first illustration of Mary Anning collecting fossils as a child, here with her father Richard. From the cover of the children's magazine *Chatterbox*, 2 November 1869.

40. Until recently it was assumed that this was Mary Anning, but despite the costume it is a sketch of William Buckland in Snowdonia in 1841 by fellow geologist Thomas Sopwith.

Cliff, some remains, which were removed to undergo an examination; the result of which is, that this specimen appears to differ widely from any which have been before discovered at Lyme, either of the Icthyosaurus or Plesiosaurus, while it approaches nearly to the structure of the Turtle. The whole osteology has not yet been satisfactorily disclosed, owing to its very recent removal. The dimensions are found to be nearly as follow: from the snout to the tail, nine feet; across the back, from the extreme points of the two front paddles, four feet; the head, which is very perfect, is not more than from four to five inches in length; the four paddles are beautifully preserved – though little has yet been done to them; the phalanges are clearly seen from the humerus to the farthest processes. It will be for the great geologists to determine by what term this creature is to be known. The great Cuvier will be informed when the bones are completely disclosed, but probably it will be christened at Oxford or London, after an account has been accurately furnished. No doubt the Directors of the British or Bristol Museums will be anxious to possess this relic of the "great Herculaneum".'[31.]

This report of the discovery suggests that the fossil was not recognised immediately as a plesiosaur. Only when she began removing the rock surrounding the bones did it become clear to Anning the true importance of what she had found. With the possible exception of Conybeare, she knew more about the skeletons of fossil marine reptiles than anyone in England and recognised that this was the long sought-after complete *Plesiosaurus*. The fossil was about 9 feet (3 m) long and 4 feet (1.2 m) wide from the tip of one paddle to the other. Its head was tiny – less than 8 inches (20 cm), but its most remarkable feature was that its neck was as long as the rest of the body. It is hard for us to imagine what Anning must have felt. This was completely unexpected. Nothing like it had been seen before.

Anning notified some of her clients and there was immediate interest in the new specimen. Within a fortnight Anning had received several offers of purchase, including one from Colonel Birch who offered her a hundred guineas despite not having seen it. On Boxing Day Anning wrote, enclosing a drawing, to fossil collector Sir Henry Bunbury whose letter expressing his interest had arrived the day before Birch's. She told him that he could have it for £110, and that she had also 'received an order from the Duke of Buckingham if not sold to send him the specimen on his account'. She recognised its significance; in her letter she writes, 'one thing I may venture to assure you it is the first and only one discovered in Europe'.[32.]

Henry De la Beche missed all the excitement. He was not in Lyme when the find was made. He had left a month earlier for what was to be a year-long visit to his family's estate in Jamaica. While there he studied the island's geology, but kept in touch with developments at home through correspondence with friends in Lyme and Bristol. A letter from Lyme told him of the find, but it was one from Conybeare in March 1824 that gave him the details.[33.]

Had De la Beche still been in Lyme he would have been the one bringing the news to Conybeare. As it was, Conybeare knew nothing of the discovery until late January 1824. He was at home in Brislington, sitting at his desk preparing a sermon to be preached in Oxford the coming Sunday, when he was interrupted by the arrival of Buckland. His friend 'brought important news – that the Annings had discovered an entire Plesiosaurus, that it had been offered to the Duke of Buckingham for £100,' Conybeare later recounted in his letter to De la Beche. Buckland had been asked by the Duke 'to call at Lyme & conclude the bargain if the Specimen was really what it purported to be.' On his way from Oxford to Lyme, Buckland had decided to drop in on Conybeare to tell him of the discovery. Conybeare asked Buckland to get him more information and three days later received 'a very fair drawing by Miss Anning of the most magnificent specimen'. Buckland must have been reassured by what he saw when he arrived in Lyme; by the end of January the newspapers were reporting that the Duke had indeed purchased the plesiosaur.[34.]

Anning's drawing was indeed 'very fair'; in fact it was such an accurate representation of the fossil that it gave Conybeare a clear idea of the whole skeleton. The drawing arrived on Friday 30 January. As there was to be a meeting of the Bristol Philosophical Society that evening, Conybeare took the opportunity to announce the discovery there. 'Such a communication could not fail to excite great interest', he told De la Beche, and 'some of the folk ran off instantly to Gutch's printing office' and Conybeare was obliged to follow to make sure the story was not distorted in the telling, which of course it subsequently was.[35.]

Having agreed the sale to the Duke of Buckingham, Buckland asked Anning to send the specimen to London where it could be examined at a meeting of the Geological Society. The saga that followed perfectly illustrates the rivalry between the geologists, Buckland and Conybeare, and Home, the anatomist. Conybeare was given the task of meeting the specimen in London 'on pain of it falling into the hands of Sir Evd. H.' Unimpressed by Sir Everard Home's work on *Ichthyosaurus*, Buckland wanted to prevent Home getting hold of *Plesiosaurus*. Buckland was also determined that news of the discovery should be broken at a meeting of the Geological Society where he was about to become President, rather than at the Royal Society by Home. Conybeare, of course, was happy to comply and after giving his sermon in Oxford, he continued to London, only to learn that the fossil had not yet reached the capital.

The easiest and safest way to transport such a specimen from Lyme to London was by sea; by wagon would have taken too long and the rough roads certain to cause damage. The Annings had had the specimen crated up, probably by Joseph, and loaded at the Cobb onto a ship, the *Unity*, which regularly plied the Lyme-London route. Unfortunately, the *Unity* ran aground on sandbanks at the mouth of the Thames and all Conybeare could do was wait. It was to be another ten days before Conybeare could see the specimen. At least he had the drawing, and he showed that at a meeting of the Royal Society Club. Charles Lyell wrote to Gideon Mantell on 17 February to tell him that

> 'W D Conybeare is in town, and has been with us for some time. He is waiting for the arrival of the new Lyme Regis Plesiosaurus of which he has an excellent drawing ... it affords a great anatomical triumph to Conybeare as most of his hypothetical restorations in his former memoir turn out true to nature.'[36.]

With many of their earlier conclusions justified by this new specimen, Conybeare told De la Beche of 'the satisfaction it afforded me to have my discoveries wh. had been questioned thus confirmed' at the Royal Society meeting, especially with Sir Everard Home present ... 'I made my Beast roar almost as loud as Buckland's Hyaenas'.

Eventually the *Unity* was refloated and made her way up the Thames, but even once the specimen had reached the Geological Society's rooms at 20 Bedford Street in Covent Garden, it was still not plain sailing. As Conybeare continued in his letter to De la Beche,

> 'at last the important package arrived & after wasting a day in vainly attempting to move it upstairs to the room of meeting at the G.S. by the aid of ten men, [we] were constrained to unpack it in the entrance passage. It is 10 feet in length and near 6 in breadth, imbedded in shale easily removed with a pen knife ... The head is remarkably small only 1/13 the whole length. The neck amazingly long, the neck and small head being equal to the body & the tail. The cervical [neck] vertebrae are 35 in number, no other quadruped having more than 9 & the swan (the longest necked of birds) only 23, but it is the number of joints rather than the total length of the neck which is remarkable for in the latter circumstance its proportions are not unlike those of the land tortoise if stript of its shell.'

In a later description of this odd creature, Buckland wrote,

> 'To the head of a Lizard, it united the teeth of a Crocodile; a neck of

enormous length, resembling the body of a Serpent; a trunk and tail having the proportions of an ordinary quadruped, the ribs of a Chameleon, and the paddles of a Whale.'[37.]

Musing on its way of life, Conybeare suggested to De la Beche that

'This same brute must have been able to nibble with his head cheek by jowl with his tail. I suppose he swam on the surface & fished with his long neck, or lurked in Shoal water hid among the weeds poking his nose to the surface to breathe & catching all the small fry that came within reach of his long sweep, but he must have kept as much as possible out of the reach of Ichthyosauri, a very junior member of whom with his long powerful jaws would have bit his neck in two without ceremony.'[38.]

With word out about the exciting new specimen, the Society's members packed into the room on the evening of Friday 20 February for a memorable meeting. As Buckland wrote to the former Prime Minister and Chancellor of Oxford University Lord Grenville five days later:

'I was in Town ... on Friday last ... for the purpose of taking the Chair at the first Meeting of the Geological Society since My Election as President, & of assisting at the Exhibition and description by Mr Conybeare of his Grace the Duke of Buckingham's magnificent Specimen of the new Plesiosaurus, & of my own Megalosaurus from Stonesfield, I was much gratified to find by far the most crowded Meeting of the Society I ever Witnessed & that a most lively and universal Interest was excited by the Business of the Evening.'[39.]

Not only did the Society's members hear Conybeare discussing the first complete *Plesiosaurus*, but Buckland, the Society's newly-installed President, followed with a paper describing some bones and teeth of a new, large, carnivorous, land reptile that had been found at Stonesfield in Oxfordshire. He named this new animal *Megalosaurus* – 'great lizard' – one of the first dinosaurs to be described, although that group would not be named or defined for another eighteen years.

When news of the discovery reached the leading comparative anatomist of the period, Georges Cuvier in Paris, he at first discounted it as being too strange to be real. Cuvier, who had himself described the skull of a large marine reptile from Maastricht in 1808, had followed the fossil discoveries in Lyme with interest. He had visited London in 1818 to meet Sir Everard Home and examine the Anning ichthyosaur skull which Home had described in 1814. When George Cumberland sent him a sketch of the plesiosaur in March 1824, Cuvier thought that it might be a forgery, a composite of two different animals either deliberately or mistakenly combined into one skeleton by Mary Anning. He wrote to Conybeare suggesting he check. Within a few days, however, Cuvier received more details and drawings from Conybeare and Buckland and was assured that this strange, long-necked animal was most certainly genuine. As a result, an impressed Cuvier subsequently held Anning in high regard, and mentions her by name in his 1824 book *Recherches sur les ossemens fossiles*.[40.]

In May 1824, the French geologist Constant Prévost arrived in England on a mission from Cuvier to purchase specimens of the new marine reptiles for the Natural History Museum in Paris to add to those already acquired at the sale of Colonel Birch's collection four years earlier. Touring the west of England with geologist Charles Lyell, Prévost was in Lyme in June 1824 to see Mary Anning and the visit is recorded in a letter from Lyell to Mantell on 9 July 1824:

> 'Three weeks since a magnificent specimen of an Ichthyosaurus (tenuirostris?) was discovered at Lyme by the celebrated Mary Anning. It was about the size of the Plesiosaurus which you saw in town. M. Prévost took a drawing of it, which I have traced, and I send to you, that you may see it, as it will be long probably ere it is published. While we were at Lyme we witnessed the discovery of a superb skeleton of Ichthyosaurus vulgaris, by Miss Anning. It was perfect, save the tail, which a cart-wheel had passed over. It was two feet long.'[41.]

Ichthyosaurs were clearly still being found by Mary Anning, but more plesiosaur fossils were also being discovered by others. The 1 February 1824 issue of *The New Monthly Magazine and Literary Journal* which had noted Anning's fossil discovery at Black Ven in December 1823 also recorded that

> 'Captain Waring, R.N., discovered, and caused to be removed from a slaty part of the blue lias ledges west of Lyme Cobb, a fine portion of organic remains, which are now arranging by that gentleman. The part which has been cleared is particularly beautiful; and there is no doubt of this specimen proving to be that of some rare antediluvian animal.'

This specimen, found within weeks of Anning's discovery by Henry Waring, a retired naval officer living in Lyme, was a partial plesiosaur skeleton, and only the second to be found. Not as complete as that found by Anning, she bought it from Waring for three pounds and, perhaps after some preparation and work, she sold it on to Prévost for ten pounds. The specimen is still in the collections of the Musée National d'Histoire Naturelle in Paris.[42.]

On his return to Paris Prévost described his acquisition of the plesiosaur and his visit to 'le petit port de Lyme-Regis', 'famous for the large number of bones of unknown reptiles' found there and how they have been brought to scientific attention by Mary Anning to the Société Philomatique de Paris. [43.]

Lyell, in his letter to Mantell, refers to 'the celebrated Mary Anning'. Her fame was now such that Anning was, as historian Hugh Torrens has put it, 'a curiosity in her own right. People now came to Lyme to visit *her*'. In June 1824, in addition to Prévost, she also received a visit from Thomas Allan, a Scottish mineralogist, and in September from Lady Harriet Silvester, widow of Sir John Silvester, a London judge. Both were struck by the extent of Anning's knowledge and by her confidence in that knowledge. Allan wrote:

> 'Mary Anning the Geologist of this Place is a very interesting person, and the scientific are entirely indebted to her for the preservation of some of the finest remains of a former world that are known in Europe. For the large specimen of Plesiosaurus, now at the Geological Society of London, and for which the Duke of Buckingham paid her £100, we are entirely indebted for her industry – and it was our good luck to find her engaged on another specimen of the same kind- equal in size, but a different animal – which she had discovered only last week ... Mary Anning's knowledge of the subject is quite surprising – she is perfectly acquainted with the anatomy of her subjects, and her account of her disputes with Buckland, whose anatomical science she holds in great contempt, was quite amusing. She walked out with us on the beach, and showed us where she looked for and found her best specimens – and in the course of our walk, she found a very fine Dorsal fin, for which I was glad to have the opportunity of giving her half-a-guinea. She says she is indebted to her father for all the knowledge she has ....'[44.]

The specimen on which Allan discovered her working was the recently-found large *Ichthyosaurus tenuirostris* noted by Charles Lyell and sketched by Constant Prévost.

Lady Harriet Silvester's view was that

> 'the extraordinary thing in this young woman is that she has made herself so thoroughly acquainted with the science that the moment she finds any bones she knows to what tribe they belong. She fixes the bones on a frame with cement and then makes drawings and has them engraved ... It is certainly a wonderful instance of divine favour – that this poor, ignorant girl should be so blessed, for by reading and application she has arrived to that degree of knowledge as to be in the habit of writing and talking with professors and other clever men on the subject, and they all acknowledge that she understands more of the science than anyone else in this kingdom.'[45.]

Mary Anning may have been poor, but she most certainly was not ignorant. Silvester uses the word in the sense of uneducated or unsophisticated; nor was she 'barely literate' as she has been described. Note also Silvester's condescending mention of 'divine favour', as if it was to divine help that Anning owed her achievements. Although her spelling sometimes tends to the phonetic, Anning's letters exhibit a strong and confident hand, even when writing in a rush to catch the post. There is no sign of uncertainty or hesitation.

Amongst Anning's papers in the Natural History Museum Library are her handwritten copies of papers published by Conybeare and others in the *Transactions of the Geological Society*. While she had received a printed copy of De la Beche and Conybeare's 1821 paper, probably given to her by De la Beche, she had not received the same courtesy from Conybeare with his 1822 paper on *Ichthyosaurus* or that of 1824 on *Plesiosaurus*. Conybeare did, however, present a copy which he inscribed to 'Miss Philpots – with the authors thanks for the information derived from her collection'; Philpot had lent him specimens and he had also examined the 'almost perfect skeleton of Plesiosaurus ... through the kind liberality of its possessor, the Duke of Buckingham.' Anning does not warrant a mention. Is this accidental or intentional? To Conybeare, she was merely the trader who supplied his Grace with the fossil and where he does refer to her it as 'the proprietor'.[46.]

Anning must have been able to borrow copies of Conybeare's papers, perhaps from Elizabeth Philpot or from De la Beche, and then set about the tedious task of copying them by hand. She copied out not just the text but carefully and skilfully drew a neat, detailed and accurate copy of each lithographed plate in ink and pencil. Without direct access or the means to purchase or subscribe to the expensive scientific literature of the day, this was her only option, even if she wanted a copy of a paper describing specimens she herself had discovered. A benefit of hand-copying, however, would be to give her a close familiarity with the contents of the papers. How she felt when she saw herself referred to as 'the proprietor' we will never know.[47.]

Allan's and Silvester's remarks give us a picture of Anning in her mid twenties as a knowledgeable, intelligent, self-assured, practical, skilled and capable woman, her confidence and knowledge allowing her, rightly, to consider herself as an equal amongst the learned professors.

A NEW FRIENDSHIP

Amongst the visitors to Lyme in the summer of 1824 was an invalid fourteen year old girl, Frances Augusta Bell, usually called Fanny, who was brought to Lyme from London by her mother and aunt in the hope that the sea air and sea-bathing would improve her condition. Born in Frome, four months after her father's death in 1809, Fanny Bell had been ill as an infant and of a delicate constitution throughout her short life, as well as suffering from poor sight. She had had little formal schooling, but 'highly gifted and adorned with a halo of piety' she took solace from the Bible. In about 1821 she developed an interest in mineralogy and in Lyme three years later, she met Mary Anning, writing to a friend that,

> 'what constituted my greatest amusement was the fossil curiosities with which Lyme abounds: and of which I made a collection, with the assistance of the young person who gains a livelihood there by so doing. Her history is so remarkable....'[48.]

Frances went on to describe how Mary Anning had been struck by lightning as a baby, and 'received no other injury but that of being turned black'; how Mary's father had died leaving the family destitute and in debt; and how Mary had sold her first fossil, sparking the idea of selling fossils to make a living. She explains that Mary

> 'has since, by her expertness in finding and selling fossils, paid off the hundred and twenty pounds debt, assisted in setting her brother up in business (that of a cabinet-maker), and now supports herself and her mother in comfortable circumstances. She is self educated; and is perfect mistress of the science of mineralogy and fossils ... She is much noticed by the ladies of the place: and such is her intelligence, that most visitors to Lyme request to be allowed to accompany her in her walks of science. But although she has raised herself so much above her rank, she retains such modesty, and humble sense of her situation, as fail not to please all who knew her.'

Fanny enjoyed Lyme; a fortnight before leaving the town in the autumn, she wrote to a friend, probably Mary Anne Davis of Frome,

> 'Lyme commands a beautiful view of the sea, as far as Portland; and is surrounded by noble cliffs, which run to a considerable distance. The air is healthy; and the principal feature of the prospect, the Cobb, a fabric of great antiquity ... projects far into the sea, and affords an agreeable promenade when Neptune is not too boisterous. You ask me if Lyme offered a good opportunity for pursuing the study of conchology; in some degree it does, but an excellent one for fossils; for the cliffs, being principally formed of blue Lias, are full of organic remains. Of both, however, I have formed a tolerable collection; and I assure you it has not been one of my least pleasing employments, endeavouring to collect duplicates of the productions for you.'

The specimens Fanny collected she sent to her friend in late October from Frome. They included part of the jaw of an ichthyosaur, along with vertebrae and part of a rib; a 'bezoar stone', which Mary Anning recognised as fossil ichthyosaur droppings, although these would not be introduced to the scientific community until later in the 1820s; ammonites; a belemnite and other fossil shells, along with some minerals such as chalcedony, selenite, calcite and quartz. The fossils were probably collected when Fanny accompanied Mary Anning on excursions to the beach.

After Fanny returned to London in mid October 1824, she and Mary Anning stayed in touch by letter, although only a few were exchanged. Despite their difference in ages, they had much in common – the loss of a father; an afflicted childhood, by poverty in Mary's case and by illness in Fanny's; both intellectually gifted and of strong religious belief; and both with an interest in geology. From the tone of the letters and the manner in which they address one another, it would seem that they had formed a close friendship during the summer.

On 23 November 1824, her fifteenth birthday, Fanny wrote to Mary:

'My very dear Mary A whole month having elapsed since I left you, and dear Lyme, without your hearing from me, I fear you will begin to think me unkind and not fulfilling my promise of writing, or suppose that illness has prevented me. My fossils ... were very much admired; particularly the metalized ammonite you gave me ... My shells also, which I have with me, traveled delightfully; and those you kindly gave me are very much admired. I assure you, that *where* we are, *there* you will not be long unknown ... I have twice visited the British Museum; and intend devoting a day to each room, in order to examine its beauties more minutely, and to improve myself in the study of minerals, shells, and fossils, as much as I can, by so good an opportunity of seeing the true specimens. O! Mary, you never saw, nor can conceive, any thing so beautiful as are the minerals. I never go there without wishing for you to participate my pleasure. They have also a good collection of shells; and fossils, in fishes and leaves, but not in the grander specimens; I mean animals and ammonites. Although the head is larger, and the eye more conspicuous, I cannot say I admire the icthyosaurus you first sent there half as much as the one you now have.'

The ichthyosaur she saw at the museum, with its large head and 'conspicuous' eye, is the first one, found by the Annings in 1811-12, by 1824 on display in the British Museum after its purchase at Bullock's sale in 1819. Fanny recognises it as the specimen collected by her friend; there would have been few other similar specimens in the museum's collection at that time, but by 1824 it was several steps removed from its discoverers, having gone from the Annings to Henley to Bullock to the British Museum. Discussing this specimen in his 1829 book, the Scottish physician and chemist Andrew Ure is unaware of its discoverers, referencing 'the fine specimen of Sir Everard Home' but not the Annings, although he does later mention other specimens 'discovered on the coast of Dorsetshire by Miss Mary Anning'. Although from time to time an author might mention that Anning found the first ichthyosaur and that the specimen was in the British Museum, as the years went by and the museum's collections grew, the fossil's identity as an Anning specimen was slowly forgotten. It was to remain so for over a century.[49.]

Fanny also compares the famous skull unfavourably with a new specimen in Anning's possession in 1824, which may be one declined by Charles Konig at the British Museum when it was offered to him the following year.[50.]

Mary Anning replied to Fanny five days later,

> 'How I envy your daily visits to the museum! Indeed I shall be greatly obliged by your sensible account of its contents; for the little information I get from the professors is one-half unintelligible ... Very little doing in the fossil world; excepting, I have found a tail for baby, and a beautiful paddle, and a few other small specimens; nothing grand or *new*.'

Mary Anning had rarely left Lyme, let alone visited the British Museum. She relied for news of geological developments away from Lyme on her acquaintance and friendship with the gentleman-geologists who bought her specimens and with whom she corresponded. Her rather sarcastic remark that she found half of what she was told by them 'unintelligible', suggests that she did not have a high opinion of some of her sources, or at least their style of communication, rather like her view of Buckland's knowledge of anatomy. In letter of 1833 to another female friend in London, she expresses her feelings of isolation from geological news:

> 'not having seen Dr. B. [Buckland] or any other professor for the last year I know no more about what is going on geology than the man in the moon.'[51.]

At the start of her reply to Fanny, acknowledging that she had not heard from her for some time, Anning writes,

> 'I have to beg your pardon for doubting your friendship; not hearing from you for six weeks instead of two, I thought if illness had been the cause of your silence, your dear good aunt would have sent me one line, just to tell me: the world has used me so unkindly, I fear it has made me suspicious of all mankind.'

What did Mary Anning mean by this last remark? The longer than expected delay in receiving a reply from Fanny seems to have upset her;

perhaps she felt that their friendship was no longer reciprocated. Her remarks suggest that this was not the first time that she had experienced this. Is she referring to some earlier disappointment in her personal life, a failed friendship or relationship, or to the difficulties she has encountered: the loss of her father, her family's poverty, the hard graft of working the shore for fossils? Or is she feeling 'used' professionally, bearing some resentment that others have made their scientific reputations on the back of her, unacknowledged, discoveries? We will never know for sure, but she makes similar comments seven years later to another young woman, so it was certainly something which she felt deeply. Perhaps she felt that she had been too generous with her knowledge and that she deserved greater recognition for her contribution to science.

The correspondence between Fanny Bell and Mary Anning lasted only six months; Fanny died on 23 May 1825, aged fifteen, less than a year after her visit to Lyme. Knowing that death was close, Fanny had arranged for small gifts to be bought and bequeathed, as memorials, to her friends. One was sent to Mary Anning who acknowledged it in a letter to Fanny's aunt in Frome on 10 August 1825:

> 'The young man delivered the parcel quite safe: my dear lamented little friend's bequest will always be revered by me. Dear kind-hearted child! amidst all her sufferings to have thought of me and my comfort: as to my remembering her, never, whilst life remains, can I forget the transient vision of her friendship. Oh! madam, had you heard her kind pious conversations with me, when we were alone, you would say that I was the most ungrateful of beings if I ever forgot her. Although so young, her mind was so heavenly gifted, that not to be doing good, to her was impossible: and I trust that, in the trials which in this world I am doomed to encounter, I shall think on her pious example, and submit without a murmur to the decrees of Providence; convinced that he only afflicts for wise purposes.'

In a later letter, of 19 October 1825, to Fanny's mother, Anning responds to a request for the return of Fanny's letters and writes movingly of Fanny as her spiritual guide:

> 'I am happy in being able in any way to gratify you; and much as I valued the letters of my lamented little angel friend, if they will afford you a melancholy pleasure, pray keep them: the contents are indelibly stamped on my memory; indeed I have reason to bless the day when first I saw dear lamented Francis. Although so many years younger than myself, she was my spiritual guide: the recollection of her pious conversations has been a support to me in the

trials I have had to sustain; it has enabled me to say, "not my will, but thine, O Lord, be done."'

In these two letters, Anning again refers to the trials she herself has had to sustain, or that she is 'doomed to encounter'.

The returned letters were published in 1827 in a memorial volume of Fanny's remarkably mature writings compiled by the Reverend Johnson Grant, the incumbent in Kentish Town from 1822, and Fanny's parish priest, who attended her in the late stages of her illness. The letters provide a rare window into the mind of Mary Anning and the strength of her religious beliefs. They also show us that she was comfortable in quickly establishing a close but brief friendship with this young woman, and trading confidences, something which she may not have had many opportunities to do.[52]

THE GREAT STORM

On the night of Fanny's fifteenth birthday, 23 November 1824, the day on which she wrote her first letter to Mary Anning, Lyme was hit by disaster. That night a hurricane-force southwesterly wind drove a storm surge that inundated much of the south coast of England from Penzance, where the lighthouse was torn apart, to Brighton where the bathing machines were smashed.

In Lyme the water overtopped the high wall of the Cobb – a wall built after the storm of 1817 – erasing about 300 feet (90 m) of it and destroying much of Cobb hamlet. The Walk, the promenade from the Assembly Rooms to the Cobb, was washed away and the damage extended to Gun Cliff. *The Western Flying Post* reported on 29 November, 'We have rarely had a more melancholy duty to perform than the recital of the tremendous effects of the gale of Monday night last. A tempest teeming with more frightful terrors is scarcely within the memory of man.' Mary Anning described the aftermath in her letter to Fanny Bell on 29 November:

> 'Oh! my dear Fanny, you cannot conceive what a scene of horror we have gone through at Lyme, in the late gale: a great part of the Cobb is demolished, every vessel and boat driven out of the harbour, and the greatest part destroyed; two of the revenue men drowned, all the back part of Mrs England's houses and yards washed down, with a greater part of the hotel, and there is not one stone left of the next house: indeed, it is quite a miracle that the inhabitants saved their lives. Every bit of the walk, from the rooms to the Cobb, is gone; and all the back parts of the houses, from the fish-market to the gun-cliff next to the baths. My brother lost, with others, a great part of his property. All the coal cellars and coals being gone, and the Cobb so shattered that no vessel

will be safe there, we shall all be obliged to sit without fires this winter: a cold prospect you will allow.'

Some modern accounts claim that the Anning's house in Cockmoile Square suffered damage, but surely Mary would have mentioned it in her letter? She does, however, state that her brother, Joseph, lost much of his property in the storm. His upholstery business was based in Bridge Street and rented from the Independent chapel's minister (and a fellow fossil dealer), the Reverend John Gleed. It was a little to the west of Cockmoile Square, on the south side of the street, and was more exposed. The Cockmoile Square house was protected on the seaward side by other buildings which took the brunt of the storm.[53.]

One of the ships driven out of the harbour was the *Unity*, the ship which had carried the Annings' plesiosaur to London earlier that year. Blown east onto the shore towards Charmouth, the crew were saved through a dramatic rescue by Captain Charles Bennett. The *Unity* was refitted at Lyme shipyard and relaunched in 1825.

A month after the storm, in December 1824, W.D. Conybeare visited Lyme with the curator of the Bristol Institution, J.S. Miller. Miller had been instructed by the Institution's Committee 'to undertake a visit to Lyme, after the great storm of November last, as directed to the augmentation of the very valuable Collection of Organic Remains, preserved in the Museum of the Institution'. In addition to being a port and the main commercial centre for the west of England, Bristol was becoming a major cultural and scientific centre. The visit was timed to take immediate advantage of any new fossil discoveries exposed by the storm, something which Anning herself would have been doing. It was a profitable journey. Miller returned to Bristol with a large collection of fossil reptile and fish specimens, some of which may have been acquired through or with the help of Mary Anning. Conybeare added to this collection with a donation of his own ichthyosaur and fish fossils from Lyme and gave an account to the Bristol Philosophical and Literary Society of their visit.[54.]

Anning's eyes were not focused solely on fossils in the Lias rocks. One day, probably in October 1825, she was walking along the shore at Charmouth at low tide when she noticed a blue clay containing horizontal pieces of blackened wood and extending under the water. She pointed this out to De la Beche who investigated further and on November he read a paper at a meeting of the Geological Society on this discovery of a submarine forest at Charmouth, crediting Anning with bringing it to his attention. Submerged forests are common features around the coast of England and Wales, dating as we now know, from the end of the last ice age as sea levels rose and drowned coastal woodlands.[55.]

THE MOVE TO BROAD STREET

On 22 July 1826, the *Sherborne Mercury* reported that

> 'Miss Mary Anning's Fossil Depot is removed to Broad Street where it will always prove a great source of attraction ... currently a splendid specimen of an Ichthyosaurus tenuirostris is on display.'

The move took Mary and her mother from the lower part of town, uphill to a little thatch-roofed shop and house on Broad Street, about halfway up on the western side of Lyme's main thoroughfare opposite George Roberts' school. Although now farther from the beach, the new shop was closer to the main centre of tourist activity and passing trade. Whether the Annings bought or, more probably, leased this property is unclear as there seems to be no evidence either way, but leasing was more common. A description from about fifteen years later has it as

> 'a plain, unpretending little shop with a small white oblong board over the door, whose primitive and crude lettering informed the public that it was richly stored with precious specimens of the saurians, pterodactyls and other fossilised remains of the denizens of primeval waters, and it was frequent resort of many of the most eminent scientists of the time.'[56.]

It functioned as Mary and Molly's home, shop and workshop where the specimens were prepared out of their rock matrix, or where larger skeletons were set in cement within a wooden frame. When the Sussex doctor and fossil collector Gideon Mantell visited in June 1832, he thought it was 'a dirty little shop, with hundreds of specimens piled around in the greatest disorder'. But Mantell was not in the best of spirits, visiting the shop after a long day in a coach from Bristol.

Lyme's streets were not numbered until the beginning of the twentieth century, and like the Cockmoile Square house, the building Anning occupied is long gone, demolished sometime after the 1930s, but it was on the site of 28 Broad Street.

SIX
Decade of Discoveries

SOON AFTER THE MOVE TO BROAD STREET, Anning found the fossil business much harder. By the mid 1820s, the initial novelty and excitement of the ichthyosaur discoveries was waning. Collectors were less willing to buy, and her fossils were not commanding the same prices as before. On top of this, her clients, both private individuals and institutions, were affected by a banking crisis in 1825 and 1826 brought about by a stock market crash and the collapse of many country banks. In the hope of increasing sales in the capital, and at De la Beche's instigation, she took on a London agent, George Brettingham Sowerby. A conchologist and natural history dealer with premises then at 156 Regent Street, G.B. Sowerby was the second son of James Sowerby, the natural history artist and publisher of *The Mineral Conchology*. On 27 July 1826, Anning wrote to Sowerby agreeing a commission of twenty per cent and describing some of the quality ichthyosaur specimens she then had for sale:

> 'On Mr De la Beaches return to Lyme yesterday, he informed me that he had been so kind as to make some arrangement with you in regard to your disposing of some first rate specimens for me, and that you Sir, had agreed to do so, at twenty percent profit, and I beg in return to say that I am willing to allow 20 per cent and that I have now by me a very perfect Skeleton of the Ichthyosaurus Communis about three feete in length, a far better than the one I sold to the Bristol Institution for fifty pounds, & I think upon the whole equal to the one I sold in Decr, last to the Geological Society its head tail paddles are in better preservation, but four vertebrae and a few of the ribs are a little out of place, the price 25 pounds, I had forgot to mention that there is part of its skin rugose or hairy – I have also a very good part of an Ichthyosaurus Intermedius in excellent preservation all but the tail which is wanting. Mr Conybeare when at Lyme offered me 15£ for it which I refused as I was not then in such distressed circumstances. I would now take twelve £ for it. This is only the second fragment of this species yet discovered – I have

also a very perfect head of the Ichthyosaurus communis two feete and half in length teeth forehead orbit of the eye all very good the price five pounds – the above specimens are fit to be placed in any collection.'[1.]

Anning was short of money; she had turned down an offer of £15 from Conybeare for a specimen she was now willing sell for £12, and for another priced at £50 she would now accept £25. Fear of a return to the grinding poverty of her adolescence haunted her still:

> 'nothing but poverty should have induced me to sell them at such low prices and I belive the will be the last I shall ever offer for sale, as I cannot discover enough get a maintenece therefor it is my intention to look out for some other pursuit.'

Her financial situation was now so dire that she seems to have been considering giving up the fossil business altogether, though what she would have done instead to earn a living remains unknown. Her letter continues,

> 'I have not any other first rate small specimens but I have a variety of secondary ones – I suppose Sir you must have heard of my large Icthyosaurus Tenuirostris which the Duke of Buckingham was to have given me hundred & ten £. pounds for it is a most splendid looking specimen twelve feete seven inches in length measuring the curve of the tail which turns up on its Back the case it is in but ... I would now take 50£ for it. The Bristol Institution would have given me seventy in april last if the Bank had not broke of course it is to large for a private collection, but if I had it in town I have not the least doubt of making by exhibition considerable more than its value.'

Anning was at the mercy of external forces. Richard Grenville, 1st Duke of Buckingham, who had purchased Anning's 1823 complete plesiosaur for somewhere between £100 and £200 – sources vary – had not bought the ichthyosaur mentioned in this letter. The widely disliked politician had been living beyond his means since before 1822 and his profligate spending had clearly come to an end (at least until he went abroad in 1827). Her second potential purchaser, the Bristol Institution, was caught up in a banking crisis in Bristol and unable to pay Anning the £70 she was happy to accept. Eventually this specimen did sell to the Bristol Institution in 1826 for £50. [2.]

Anning sent several of the specimens listed in her letter – the well-preserved three-foot (one metre) *Ichthyosaurus communis* skeleton and the head of the same species – to Sowerby in August 1826. Two months later they remained unsold. Anning wrote to him again in early October, clearly still in need of funds:

> 'I have taken the liberty of writing to ask you if there is any prospect of your soon disposing of the small ichthyosaurus and head which I sent the first week in August last. I shall not have troubled you so soon if necessity did not oblige me. I believe Sir that I mentioned in my last that if you liked me to send you any other specimens from this neighbourhood that I should be very happy to send some of the best I can get as the winter is the best time for collecting them.'[3.]

Her letter prompted Sowerby to take action. He offered the ichthyosaur skeleton to Adam Sedgwick in Cambridge, but Sedgwick thought it too expensive. In a letter to Sowerby of 30 October 1826, Sedgwick adds as a postscript, 'You offered me the refusal of the Ichthyosaurus. If any one offers the money take it without waiting for my answer. The price is I think too large ...'[4.]

Anning's arrangement with Sowerby lasted at least until 1829, although at times she had difficulty getting paid. She wrote to Sowerby in March 1829 asking for payment due to her and requested that it be passed to her through John Philpot, the London-based solicitor and brother of her friend Elizabeth Philpot:

> 'Mary Anning will thank Mr Sowerby to pay to Mr Philpot the ten pounds now, and give Mr Philpot a Bill for the remainder at two months or sooner if possible as she really wants the money.'[5.]

Two months later, and she has to chase Sowerby for the balance due to her:

> 'Mary Anning takes the liberty of reminding Mr Sowerby that the two months expired on the first of May and that she would be obliged to Mr Sowerby to pay to Mr Philpot the remainder of the bill due to her, viz £13.2.5 as she really wants the money.'[6.]

This finally spurred Sowerby to pay up, so on 15 May:

> 'Mary Anning begs to acknowledge the receipt of the residue of the money due to her viz thirteen pounds two shillings & five pence, having received ten pounds in March last making in whole £23.2s.5d being the amount due to her for fossils.
>
> Mary Anning has not any fine Specimens of an Ichthyosaurus by her at present.'[7.]

One cannot but feel sorry for Anning in this period of the mid to late 1820s; she was short of really good specimens, her London agent was slow to sell what she sent him, and then delayed forwarding the money she so desperately needed following a successful sale. The lax attitude to the timely payment of bills she encountered from clients and institutions much wealthier than her may have contributed to the rather caustic view of the world and human foibles she occasionally expressed.

AMERICA COMES TO LYME

In October 1827, Anning had a visit from a larger than life and rather remarkable American, George William Featherstonhaugh. Born in London, Featherstonhaugh had been brought up in Scarborough where he developed an interest in geology and natural history. He moved to the United States in 1806, becoming a farmer in New York State and pioneering the development of steam railroads there, the first of which was completed in 1831. Elected a Fellow of the Geological Society in 1827, in 1834 Featherstonhaugh was appointed the first geologist to work for the US Government with the task of exploring part of the huge tract of territory acquired by the United States in the Louisiana Purchase of 1803, much of which he did by canoe.

On a return visit to England in 1826-27, Featherstonhaugh made the acquaintance of the geologist William Smith in Scarborough and, in London, sought out the coterie of geologists at the heart of the Geological Society. He visited Mary Anning's dealer, G.B. Sowerby, buying 1,750 specimens, largely from the collection of James Parkinson, a founder member of the Geological Society. Following a visit to Devon with Buckland, Featherstonhaugh called at Lyme at Buckland's suggestion to add some Lias fossils to his collection. He spent a Sunday morning with Mary Anning and was much taken with her, calling her a 'very clever funny creature'. Big hearted, free of prejudice, a fearless adventurer, Featherstonhaugh was very different to most of Mary's visitors, and it seems likely that she was able to relax and talk freely in his company. Making his way back from Devon separately, Buckland also visited Anning's shop where he purchased more specimens for Featherstonhaugh and found her hard at work excavating a large mass of a Lias crinoid for the American.[8.]

Although they superficially resemble plants, with a root-like base and a long flexible stem ending in many branches, and are often called 'sea-lilies', crinoids are marine animals related to starfish and sea-urchins, the group called echinoderms. Basically, crinoids look like a starfish on a stalk. They are common fossils through a long period of geological time and are still around today, although now living in deep waters. Like all echinoderms, crinoids are made up of many small plates of calcite tightly fitted together. The main part of the animal lives in a 'cup' at the end of the stem surrounded by branching arms which collect and direct food-particles down to the crinoid's mouth.

When the animal dies, the crinoid usually breaks up into its many separate plates. Some of the most robust of these are the segments which make up the stem. In most crinoids these segments, called ossicles, are circular in cross-section, like mints with a hole in the middle, but in the Lias rocks of Dorset, they usually have five-sided, sometimes star-shaped, ossicles, hence the name *Pentacrinites*.

Anning was probably extracting Featherstonhaugh's specimen on the side of Black Ven close to Charmouth and from the sequence of rocks which geologists today call the Black Ven Marls. Within the marls is a band of rock a couple of metres thick in which *Pentacrinites* occurs, often in spectacular masses of near-complete fossils. This crinoid was first named by a German naturalist, Johann Friedrich Blumenbach in 1804, who based his description on a specimen from Dorset, most likely from the same site and stratum from which Anning was now collecting. It is still a source of *Pentacrinites* to local fossil collectors today. It proved a profitable sequence of rocks for her as she sold specimens of *Pentacrinites* to several collectors and museums. Although these fossils were not as valuable as the marine reptiles – at the sale of Colonel Birch's collection in 1820, slabs of *Pentacrinites* sold for between five shillings and three pounds – large slabs could fetch a good price. One was sold in 1837 for £16 6s to John Templeman of Bath who had a summer house in Lyme.[9]

Featherstonhaugh returned to New York with some eight thousand specimens of minerals and fossils which included fossils from Yorkshire, from Gideon Mantell in Sussex and from Mary Anning in Lyme, all of which he deposited in the New York Lyceum of Natural History. Sadly, the Lyceum's collections were destroyed in a fire in 1866.[10]

Featherstonhaugh's visit to Lyme may have brought her further orders for marine reptiles from the United States. In December 1827, the newspapers were reporting:

> 'Miss Mary Anning, of Lime, has several orders for the grand specimens of the Lizard tribe to be sent to America for Museums there.'[11]

Notwithstanding her financial worries and her thoughts of seeking an alternative occupation, the late 1820s were a particularly productive period for Mary Anning. By 1828 she had several ichthyosaur specimens for sale in her shop. In late November the press were reporting that 'Miss Anning, of Lyme, has found a large perfect specimen of the Dapedium Politum, an antediluvian fish with oblong scales'. This she sold to John Templeman. Although this *Dapedium* was a well-preserved example, it was not a new species. But some of her finds were new and scientifically important discoveries, although most would not necessarily bring her any significant income. Amongst these were fossilised squid ink and ichthyosaur faeces.[12.]

SEPIA

In 1828, Anning collected small, brown-black lumps from the grey mudstones that form the upper part of Black Ven. The rocks there are known to geologists as the Belemnite Marl for the abundance of belemnite fossils that they contain. Belemnites are extinct cephalopods which in life probably resembled today's squid or cuttlefish and, like the latter, had a solid internal shell which is the part most commonly preserved as a fossil. Called the 'rostrum' or 'guard', these are narrow, elongate conical fossils of dark grey-brown calcite which taper to a bullet-shaped point. Broken sections of belemnite guards are readily found along the shore at the foot of Black Ven. They had been known for centuries from various parts of England and had entered regional folklore as 'thunderbolts', darts thrown down from heaven during thunderstorms. In fact the name, belemnite, is derived from the Greek, *belemnon*, meaning 'dart'. The belemnite guard probably acted as a counterweight to the animal's body and arms. Belemnites would have been amongst the first fossils gathered by Mary as a girl and she would undoubtedly have stocked examples for sale in her shop.

In a letter to William Buckland of November 1834, she describes finding two rock layers at Golden Cap east of Charmouth packed with 'imense numbers' of belemnite fossils, some with other shells attached to them. She recognised that this showed that the rocks had been 'quietly deposited' and that the belemnites had lain on the sea floor for some time, long enough for their soft tissues to rot and for other shells to settle and grow on the belemnite guards. Anning was not just collecting fossils, but interpreting how they formed from the evidence she saw in the rocks. Her observations described in this letter found their way straight into Buckland's *Bridgewater Treatise* which he was then preparing for publication.[13.]

On 6 February 1829, at a meeting of the Geological Society, now in new premises at Somerset House, Buckland read a wide-ranging paper, summarising some recent remarkable discoveries by Mary Anning in the Lias of Lyme Regis which he elaborated upon in several later publications. The small brown-black masses, he surmised, were probably the fossilised ink-sacs of an unknown squid-like animal related to belemnites but lacking the solid internal shell. Squid today, when threatened, produce a cloud of dark ink to mask their escape and it would appear that their Jurassic relatives did the same. On a visit to Lyme in 1829, Buckland had dissected a modern cuttlefish and examined its ink-sac and noted its similarities to the fossils.

Buckland was astonished that 'traces of so delicate a fluid as the ink which was contained within the bodies of extinct species of Cephalopods' could have survived, yet clearly it had. Buckland recognised that this could only have happened if the animal had died suddenly and been buried quickly by sediment on the sea floor. He readily gave credit to Anning: 'we owe this discovery', he wrote in 1836, 'to the industry and skill of Miss Mary Anning, to whom the scientific world is largely indebted, for having brought to light so many interesting remains of fossil Reptiles from the Lias at Lyme Regis'.

Squid ink – sepia – was a common drawing medium much used by artists. In 1826, Buckland sent a sample of this fossil sepia to his friend, the sculptor Sir Francis Chantrey, suggesting that he try it as a pigment. This he did, and used it in a drawing which he showed to 'a celebrated painter, without any information as to its origin, and he immediately pronounced it to be tinted with sepia of excellent quality, and begged to be informed by what colourman it was prepared'. The melanin which gives cephalopod ink its colour was unaltered after 200 million years.[14.]

Elizabeth Philpot also tried her hand at drawing with the fossil sepia. In December 1833, she wrote to Mary Buckland in Oxford, enclosing a sepia drawing of an ichthyosaur skull,

> 'I fear you will think of me presuming in offering you a sketch of a fossil which was made to unskilful a hand as mine; but I am induced to send it to you as it was drawn with the colour prepared from the Ink of the fossil Sepia; and Mary Anning considers this Jaw the most perfect perfect [sic] specimen she has ever met with.'[15.]

Both Buckland and his wife wrote back separately to Philpot on 1 January 1834, impressed not only by the sepia drawing itself but also by the quality of the specimen. Mary Buckland called it 'quite a Bijou as a specimen, even among your Collection of beauties' while Buckland himself wished that he had seen it sooner as he would 'have craved permission to engrave in my Book now soon forthcoming, instead of a drawing now in Preparation of the Head in the British Museum which Sir E. Home engraved in the Philosophical Transactions. I think your specimen must shew points of anatomy which the other does not …'. In reply, Philpot sent Buckland a more detailed drawing of the same ichthyosaur skull, at its natural size, but the main part of their correspondence dealt with fossil ink-sacs and whether they had been found attached to belemnites.[16.]

In 1829 Buckland had hesitated to link these fossil ink-sacs to belemnites in the absence of a specimen which showed the connection, although he said that Mary Anning considered that this was the case. In 1834, the Swiss geologist Louis Agassiz went to Lyme to examine the Philpot collection, recognised two examples and duly notified Buckland. In 1836 Buckland named the fossils *Belemnosepia*.[17]

Perhaps inspired by Buckland, Mary Anning took the interesting step of finding a modern animal which produces ink, probably a sea hare, the gastropod *Aplysia punctata*, and took it to Morley Cottage where she and Elizabeth Philpot dissected it. She described her experiment in a letter to William Buckland on 21 December 1830:

'Do you reccolect the live creatures you had to put in spirit (if so my discription may be acceptable) I got two more, one alive, and it, whenever it was touched injected a purple fluid (this one Mr. De la Beach coveted, and has taken to some Naturalist to be described). The second I dissected at Miss Philpot's, it first had a shell this shape [sketch] very like a smooth pecten only more concave, also a sack or ink bag exactly resembling the fossil one you borrowed of Miss Philpot this shape [sketch] it also had a second small bag (like the gizard of a fowel) containing a number of horny triangular little pieces (I should think to assist its digestion) it had four horns, in the two uppermost are its eyes; round the shell, I should have observed, a lip of fleshy substance which completely covers and gards it in case of necessity.'[18]

Here Anning is conducting a piece of investigative science, demonstrating that far from being a mere collector and supplier of fossils, she attempted to understand what she was finding through practical experiment and by reading whatever scientific literature she could lay her hands on – even laboriously copying it out.

COPROLITES

'Miss Mary Anning has collected a treat for the fossilist, and admirers of the wonders of an antediluvian world, in three beautiful specimens lately found: one an Icthyosaurus Intermedius, two Ichthyosauri Vulgares. Dr Buckland will be much pleased at a confirmation of his opinions respecting the excrement of those and other animals which has been by some considered to be the Bezoar stone. Miss Anning has found a Cornua Ammonis of unusual size and beauty.'[19]

The New Monthly Magazine and Literary Journal, 1 July 1828.

Buckland's February 1829 paper read at the Geological Society also dealt with other rounded, dark grey, twisted rock lumps found in the Lias rocks at Lyme. These were called 'bezoar stones' because of their resemblance to the stony concretions of undigested food expelled by ruminants such as cattle and goats, the name coming from the bezoar ibex of the Middle East. Traditionally these stones had been considered to be antidotes to poison. Bezoar stones, Buckland announced to his astounded audience at the Geological Society, were the fossil faeces of ichthyosaurs and contained the undigested bones and scales of fish. That the specimens from Lyme were ichthyosaur droppings was suggested by specimens found by Mary Anning and he credited her with this in his published paper:

> 'The certainty of the origin I am now assigning to these Coprolites, is established by their frequent presence in the abdominal region of the numerous small skeletons of Ichthyosauri, which, together with many large skeletons of Ichthyosauri and Plesiosauri, have been found in the cliffs at Lyme, and supplied to various collectors by the skill and industry of Miss Mary Anning. I have two of these skeletons, in each of which the Coprolites are very apparent, but flattened; and Miss Anning informs me that since her attention has been directed to these bodies, she has found them within the ribs or near the pelvis of almost every perfect skeleton of Ichthyosaurus which she has discovered.'[20]

Through this common association, Anning had realised some years earlier, in 1824, that bezoar stones were fossilised droppings. In December 1825 Buckland sent a specimen for analysis to the mineralogist and chemist William Wollaston who showed it to be rich in calcium phosphate, and hence derived from bone. Such phosphate-rich lumps were familiar to Buckland. In 1821 he had excavated a cave at Kirkdale in Yorkshire containing broken, chewed bones and white nodules which he identified as hyaena bones and droppings. Buckland also had specimens from Lyme analysed by the chemist and physician William Prout who attributed the dark colouring of the stones to the sepia ink of fossil squid, suggesting that, in addition to fish, these animals were also ichthyosaur prey.[21]

Buckland coined the term 'coprolite' – Greek for 'dung stone' – for these fossil faeces which, he noted, occur over a wide range of strata. With these odd stones now identified, geologists around the country sent Buckland specimens from different regions and rock sequences. Noticing that some of the Lyme Regis coprolites had a spiral structure, Buckland dissected modern rays and dogfish and injected their intestines with a quick-setting cement to see if he could replicate the same spiral form. Buckland was a pioneer of this kind of experimental palaeontology; following his discovery of broken and chewed bones at Kirkdale Cavern in Yorkshire in 1821, he fed ox bones to hyaenas in a travelling menagerie visiting Oxford and found that the bite marks inflicted by the modern animals matched those on the fossil bones and that the droppings they produced as a result were the same as those he had found in Kirkdale Cavern.[22]

Coprolites became something of an obsession with Buckland who was noted for his rather coarse sense of humour, and his friends and colleagues pandered to his coprolitic predilection with verses and sketches:

'The noble science of Geology
Is bottomed firmly on Coprology' [23.]

At a meeting of the British Association for the Advancement of Science in Edinburgh in 1834, Buckland visited the shore of the Firth of Forth where he collected a large number of rounded nodules containing fish coprolites: 'a few minutes sufficed to collect more specimens than I could carry'. These stones were known locally as 'beetle stones' and were used, cut and polished, to make table tops and ornaments by Edinburgh lapidaries. Taken by this idea, Buckland either purchased or had made a table whose top was inlaid with sixty-four of these cut and polished fish coprolite nodules. This was kept in Buckland's drawing room in Oxford and later at the Deanery in Westminster, and, according to his son Frank, it 'was often much admired by persons who had not the least idea of what they were looking at.' Today it is much admired by visitors to Lyme Regis Museum to which the table was donated by Buckland's grandson in 1928.[24.]

Anning's discovery and recognition of coprolites helped Buckland put their study on a scientific footing. The discoveries were hers, but it was Buckland who recognised them as a key to learning about feeding and predation in the distant past:[25.]

> 'When we see the body of an Ichthyosaurus, still containing the food it had eaten just before its death, and its ribs still surrounding the remains of fishes, that were swallowed ten thousand, or more than ten times ten thousand years ago, all these vast intervals seem annihilated, time altogether disappears, and we are almost brought into as immediate contact with events of immeasurably distant periods, as with the affairs of yesterday.'[26.]

PTEROSAURS

'This specimen of Pterodactyle was discovered, in December last, by Miss Mary Anning, and was found to belong to a new species of that extinct genus.'[27.]

Anning's good fortune in the late 1820s continued with another remarkable fossil reptile discovery, not a marine reptile this time, but a flying one. This was the first recognised example of a pterosaur – called at the time a pterodactyle – found in Britain, which Anning discovered in December 1828.

It was not the first ever found; examples had been known for many years from limestones in Bavaria, but now flying reptiles could be added to the list of strange animals from the Lias rocks of Lyme Regis. The precise location of its discovery is not recorded but it most probably came from Church Cliffs. Buckland announced the discovery in his February 1829 Geological Society lecture. The specimen was incomplete, lacking the skull, and although the bones were displaced, most of the body and limbs were present. Its long claws, different from other specimens known, suggested that it was a new species which Buckland called *Pterodactylus macronyx*, although initially he had proposed naming it *Pterodactylus Dorsetiensis*.

Buckland had long suspected that pterosaur bones were present in the Lyme rocks as he had heard about, but never saw, 'a skeleton of a fossil bird' belonging to Charmouth fossil collector William Rowe. This was about twenty years before, so around 1809. Following Mary Anning's discovery, he thought it likely that this earlier specimen may have been a pterosaur. He had also seen, in the collection of the Misses Philpot at Morley Cottage, in 1825, some wing and toe bones that he could not then identify and, in more recent years, a 'jaw of a Pterodactyle' in the Philpot collection. He also now realised that bird bones that had been reported from rocks in Oxfordshire and Sussex were in fact pterosaur remains. This new specimen was, however, the most complete, even if its skull was absent. The length of the bones suggested to Buckland that Anning's pterosaur was about the size of a raven, with a wingspan of over a metre.[28.]

Buckland painted a vivid picture of the animal in the Lias world:

'In size and general form and in the disposition and character of its wings, this fossil genus ... somewhat resembled our modern bats and vampyres, but had its beak elongated like the bill of a woodcock, and armed with teeth like the snout of a crocodile; its vertebrae, ribs, pelvis, legs and feet, resembled those of a lizard; its three anterior fingers terminated in long hooked claws like that on the fore-finger of the bat; and over its body was a covering, neither composed of feathers as in the bird, nor of hair as in the bat, but of scaly armour like that of an Iguana; – in short, a monster resembling nothing that has ever been seen or heard-of upon the earth, excepting the dragons of romance and heraldry. Moreover, it was probably noctivagous and insectivorous, and in both these points resembled the bat; but differed from it, in having the most important bones in its body instructed after the manner of those reptiles. With flocks of such-like creatures flying in the air, and shoals of no less monstrous Ichthyosauri and Plesiosauri swarming the primeval lakes and rivers, – air, sea, and land must have been strangely tenanted in these early periods of our infant world.'[29.]

Buckland's description of the heraldic dragon inspired a fanciful reconstruction in a dark, almost monochrome, painting of 1829, the reverse of which is inscribed, 'By the Revd. G. Howman from Dr Burckhardt's [Buckland's] account of a flying Dragon found at Lyme Regis supposed to be noctivagous'. This shows a nocturnal scene of a flying dragon, with a long, curled, serpentine, pointed tail and bat-like wings, above a rocky cliff on which stands a ruined castle, while offshore a sailing ship lists heavily in a stormy sea. The sinister, almost nightmarish, quality of the scene is diminished somewhat by the dragon who would not look out of place in Oliver Postgate's *Noggin the Nog*.[30.]

In 1829 when the picture was painted, the Reverend George Ernest Howman was vicar of Sonning in Berkshire. He had an interest in geology and as a student at Oxford he may have attended Buckland's inaugural address as Reader in Geology in 1819. Buckland's justification of geological knowledge with Biblical accounts of the creation would have found a willing disciple in Howman. How he came to hear of Buckland's description of the pterosaur is not clear, but his parish was only 30 miles (50 km) from Oxford. He was certainly a competent artist, and at least one other picture of his, a watercolour view of Babbacombe in Devon, is known. Howman's noctivagous dragon picture is now in Lyme Regis Museum, having been acquired from Lady Isobel Kerr, granddaughter of John Kerr, 7th Marquis of Lothian to whom Howman was domestic chaplain from 1824.[31.]

News of the discovery had reached Paris by February 1829, when Charles Lyell, attending a lecture by Constant Prévost, with whom he had been in Lyme five years earlier, 'announced Mary Anning's new Pterodactyle of Lyme' to those present.[32.]

Even though Anning's specimen was the first flying reptile to be identified in Britain, the British Museum was reluctant to purchase it and Buckland bought it himself from Anning. It was eventually acquired by the Museum in 1835. Further pterosaur material, two skulls, were found at Lyme, after Anning's death, in the 1850s and 1860s. Both were acquired by the British Museum and described by Richard Owen in 1859. As the skulls differed from those of the German *Pterodactylus*, Owen renamed the Anning-Buckland specimen as a new genus, *Dimorphodon macronyx*.[33.]

CRINOIDS AND FOSSIL WOOD

A much less well-known discovery made by Anning at about this time was that the common fossil crinoid, *Pentacrinites*, was often found associated with pieces of fossil wood or lignite. In a letter to Buckland in November 1834 she tells him that 'the pentacrinite lies with the head downwards invariably, with the footstalks uppermost on which lies the lignite'. In his 1836 book, *Geology and Mineralogy considered with reference to Natural Theology*, Buckland gave a detailed account of what 'Miss Anning has constantly observed in this Lignite', her observations leading to the conclusion that

> 'the creatures had attached themselves, in large groups, (like modern barnacles), to the masses of floating wood, which, together with them, were suddenly buried in the mud, whose accumulation gave origin to the marl.'

Anning's long experience of the fossils of the Lias provided the evidence that this crinoid lived, not fixed to the sea floor as many do, but floating attached to pieces of driftwood.[34.]

A PERFECT PLESIOSAUR

On Thursday 29 January 1829 Mary Anning found another complete plesiosaur, the same species as the one of 1823, but bigger and in better condition: 'Miss Mary Anning has just discovered another specimen of the Plesiosaurus dolichodeirus ... eleven feet in length, and nearly perfect', announced the *Dorset County Chronicle*. Anning wrote to Buckland on 9 February 1829 to inform him of her discovery and to ask his opinion on what she should do with the specimen. She was concerned that Charles Konig at the British Museum might have taken offence at not having been given first refusal and asked Buckland to intervene as she thought it should stay in England:

> 'I embrace the first oppertunity of informing you that I have discovered another plesiosaurus superior to the one purchased by the Duke of Buckingham. The head is really beautiful and the lower jaw has sliped from under the upper jaw by which we can see the inside of the mouth. The Creature is between eleven and twelve feete in length and four broade Sir Mr König seemed offendid at my not offering him the first. Therefore I will ask you to mention it ... Four Museums have at differant periods bespoke a plesiosaurs namely Bristol Institution, British Museum, Paris Mr Featherston for America but I hope it will be purchased in England as it is really Magnificent ... I have not yet washed it but from what I can see there will be traces of a skin or shell between the bones. I found it Thursday 29 Jan and have been ever since setting and picking it ... Sir I shall feel greatly obliged by your sending me a line to say what you think I had best do in regard to disposing of it. I must write to the Bristol Institution to say I have got such a thing.'[35.]

A week later she wrote again to Buckland in reply to a letter of his asking for a drawing of the specimen:

'Sir I did not receive your letter time to send any sort of sketch, there will not be two hours daylight until the post goes out again, & if there was time my eyes are so inflamed with picking it that I could not see to draw it but Sir you may take my word when I say it is far more perfect than the one the Duke purchased.'

Anning had been working on preparing – 'picking' – the fossil, the painstaking work of carefully removing the rock to expose the bones, for over a fortnight. Her eyes were too tired to draw, especially as the short February day was drawing to a close.[36.]

Buckland passed Anning's two letters to Konig at the British Museum, encouraging the museum to purchase not only the plesiosaur for one hundred guineas but also the pterosaur for £20. To add a little pressure he tells Konig that 'M.A. has a Commission to send it [the plesiosaur] to Cuvier in case the Museum shd decline to take it'.[37.]

There was interest, too, from the United States in acquiring this specimen for the Academy of Natural Sciences in Philadelphia, but Buckland successfully persuaded a reluctant Konig to buy it for the British Museum. In a letter to Anning on 1 April 1829, Konig hoped that her description of the specimen 'may not prove exaggerated & that the specimen be really better superior to that of the Duke of Buckingham ... Dr Buckland would be greatly terribly vexed should ... there be any disappointment'.[38.]

Its acquisition was announced in May 1829:

'The grand specimen of the *Plesiosaurus*, unequalled in any country, now at Lyme, where it was discovered by Miss Mary Anning, is not destined for the Museum of Philadelphia. It has been purchased by the trustees of the British Museum at a cost of 100 guineas.'[39.]

By the late 1820s, Mary Anning's reputation was at its height. She was respected as a reliable source of high quality fossils which had been carefully excavated, prepared and mounted. She was an observant, knowledgeable collector who had first brought to light the strange animals that had inhabited Dorset in the distant past – the ichthyosaurs, plesiosaurs and pterosaurs – as well as shedding light on the odd bezoar stones, the brown masses of fossil sepia and the mode of life of *Pentacrinites*. Her specimens were in the collections of museums and gentlemen around England and further afield, in France and New York. She was highly regarded by the great French comparative anatomist Georges Cuvier. And she had never left Lyme.

MARY MEETS THE MURCHISONS

What little we know of Mary Anning's life suggests that she left Lyme for an extended time on only a handful of occasions. Her first, and perhaps farthest, journey was one undertaken in July 1829 when, at the invitation of Charlotte and Roderick Murchison, she visited their London home.

Anning first met the Murchisons in Lyme in September 1825 when they were on a geological tour of the south coast of England. When they reached Lyme, the 36 year old Charlotte was fatigued – she had contracted malaria while travelling in Italy in 1817 – so her husband left her there to, he wrote,

> 'amuse herself, and become a good practical fossilist, by working with the celebrated Mary Anning of that place, and trudging with her (pattens on their feet) along the shore; and thus my first collection was much enriched.'

Like the fossil hunts led by local collectors for visitors to Lyme today, heading out on a fossil excursion along the beach with Mary Anning seems to have been an essential activity for well-to-do visitors to Lyme and, at least in some cases, something which she did without a fee. [40.]

Roderick Impey Murchison had come late to geology. Born into a family of Scottish Highland landowners in 1792, he joined the army, seeing action in Portugal and Spain with Sir Arthur Wellesley – later the Duke of Wellington – during the Peninsular War of 1807-1814 and serving at the Battle of Corunna in 1809 with General Sir John Moore. With the ending of the Napoleonic Wars at the Battle of Waterloo in June 1815, Murchison, like many other officers, was surplus to the army's peacetime requirements and retired on half-pay. That August, he married Charlotte Hugonin, a general's daughter. Her father was a keen astronomer, her mother a botanist, so Charlotte was brought up with a keen interest in the natural world. Murchison described her as 'an attractive, piquante, clever, highly educated young lady'. She was also a skilled landscape artist who had been taught by the leading watercolour artist of the day, Paul Sandby.[41.]

During the early years of their married life, Charlotte used her undoubted charms to redirect her husband's interests away from foxhunting in favour of more cultivated and intellectual pursuits. The mathematician and astronomer Mary Somerville whom the Murchisons met in Rome in 1817 recalled that

> 'Mrs Murchison was an amiable accomplished woman, drew prettily & what was rare at the time she had studied science, especially geology and it was chiefly owing to her example that her husband turned his mind to those pursuits in which he afterwards obtained such distinction.'

Mr Murchison, however, at that time 'hardly knew one stone from another'. [42.]

A chance meeting with Sir Humphrey Davy, President of the Royal Society, in the summer of 1823 led to a move to London and an opportunity for Roderick to pursue his new interest in science, and geology in particular, which he took to with gusto. He joined the Geological Society, becoming its President in 1831. He was at his happiest doing fieldwork, often accompanied by Charlotte. His examination of the rocks of South Wales and the Welsh borders eventually led, in 1839, to his defining and naming a new unit of the rock sequence of Britain, the Silurian System. In 1836, with Adam Sedgwick, he resolved the position of a problematic sequence of rocks in Devon, recognising them as a unit which they termed the Devonian System. Murchison's ordering of the rock sequence did not end with the Silurian and Devonian Systems, nor was his work confined to Britain. In the early 1840s, Murchison's researches in Russia led him to define the Permian System of rocks, named after the town of Perm in the Ural Mountains. The terms Silurian, Devonian and Permian are still used today for those rocks and for periods of geological time.[43.]

Murchison's interests extended beyond geology; he was one of the founders of the Royal Geographical Society in 1830, and served four terms as its President. He was knighted in 1846 and in 1855 succeeded De la Beche as Director-General of the Geological Survey, a post he held until his death in 1871. Murchison was one of the towering figures of nineteenth century geology and geography, and mountains, rivers, and waterfalls on five continents bear his name – as does a crater on the moon.

In 1825, Murchison was but a novice, and his wife the more experienced geologist. Charlotte's time learning to become 'a good practical fossilist' with Mary Anning was well-spent; the following year, when she and her husband were touring Yorkshire and Scotland, Charlotte collected fossils, some of which she sent to James De Carle Sowerby for inclusion in *The Mineral Conchology of Great Britain*, the publication which he continued after the death of his father, James Sowerby. An ammonite which Charlotte collected near Portree on the Isle of Skye was named after her (not her husband) by Sowerby, *Ammonites murchisonae* (now called *Ludwigia murchisonae*).

LEAVING LYME FOR LONDON

Mary Anning and Charlotte Murchison became friends during that 1825 visit, and corresponded subsequently. In an undated letter to Charlotte, but probably written in late 1828, Anning acknowledged an invitation to visit London: 'I cannot express how gratefully I am obliged for your kind invitation to town I think should anything occur to prevent my accepting it it will be the death of me'. In a letter the following February, Anning is not yet able to confirm when she might visit:

> 'I have defered answering your kind letter to the last moment in hope that I should have been able to have accepted of your kind invitation, but I am obliged to say that I cannot do so. To say how much I am disappointed would be needless, independant of the honour of being invited to your house I have never been out of the smoke of Lyme – as my journey to London depended on a letter and I have not yet received it I no longer hope to receive by the time specified in yours I can truly say that hope deffered maketh the heart sick.'[44.]

Anning's remarks here, in tandem with her earlier comments to Fanny Bell about Fanny's visit to the British Museum, imply a yearning for wider horizons, or at least a trip to London, the centre of the intellectual debates to which her discoveries had contributed so much and from which she was excluded by poverty and geography, as well as by gender. She explained to Charlotte that the postponement was due to waiting for a letter. Perhaps she was expecting news of a sale, perhaps even confirmation of purchase of the plesiosaur she had just discovered. Its sale for 100 guineas would certainly have provided the funds she needed, though it is possible that the Murchisons may have offered to pay for her travel.

Eventually, and probably with mixed excitement and apprehension, aged thirty, she left Lyme for the first time in her life, in the summer of 1829. We have no record of how she travelled. It may have been by coach which left shortly after dawn from the George Inn at nearby Axminster, depositing its weary passengers outside the Swan with Two Necks in Cheapside 18 hours later, or it may have been by sea, the route taken by her specimens. The *Commerce*, the *Honiton Packet* and the *Unity* – relaunched after being damaged in the Great Storm of 1824 – sailed every fortnight for London. We also don't know where she stayed in London, but the letter implies that it was at the Murchisons' house, 3 Bryanston Place. Roderick Murchison was not at home at the time of her visit as he had left on a tour of the Eastern Alps with Adam Sedgwick. That she was Charlotte's house guest tells us much about how Anning was perceived: as a friend and an equal, not as a working class trader.

A few pages of a journal of Anning's visit survive to tell us a little of what she saw and did on her visit to London. At the Geological Society's rooms at Somerset House she met the recently appointed Curator and Librarian, William Lonsdale, a 'tall, grave man' according to Roderick Murchison, who showed her the Society's collections. On seeing these, she was struck by the resemblance of fossil fish from rocks which we now know to be about 50 million years old to those with which she was familiar in the Lias rocks of Lyme. She was also shown a cast of one of her specimens of *Plesiosaurus* which had been sent to Cuvier in Paris, 'so like that I could hardly distinguish the difference', she wrote. She went to the 'Zoological rooms' where she saw 'the skeleton of a bat some of the bones much resembling my pterodactyle'.

Her use of the possessive pronoun here may suggest her feeling of ownership of her discoveries, or perhaps simply that that particular specimen was, at that stage, still hers, and had not yet been purchased by Buckland. It is not clear from her notes where she saw the bat skeleton. The 'zoological rooms' seem not to have been the British Museum which she visited two days later; she may have been referring to the Hunterian Museum of the Royal College of Surgeons where some of her ichthyosaur specimens resided, or, perhaps more likely, she saw it in the zoological collection of Dr Robert Grant, first Professor of Zoology and Comparative Anatomy at the newly established University of London.[45.]

She was impressed by the newly-constructed Regent Street and its centrepiece, The Quadrant, and by the exhibits at the British Museum which she visited two days after the Geological Society. Curiously, her surviving notes make no mention of the fossils at the Museum but she was delighted by what she saw in the King's Library – noting particularly the prayer books of King Ethelbert, Lady Jane Grey, and King Henry VII – and by the Elgin marbles and Egyptian sculpture and mummies. Anning also attended a service at St Mary's Church which had been constructed just five years earlier, on Bryanstone Square close to Murchison's home, where she heard the rector and well-known bibliographer, Thomas Frognall Dibdin preach. She writes, too, simply, that she went to 'Mr Sowerby's museum'. This was presumably a visit to the Regent Street premises of her dealer George Brettingham Sowerby, perhaps to chase him for money owed. One wonders if Sowerby was expecting her or if this was an unannounced visit intended to take him by surprise.[46.]

GREENSAND MANIA

In her letters to Charlotte Murchison, Anning tells her of some of her recent collecting activity with Elizabeth Philpot. Encouraged by Charlotte and De la Beche, she had turned her attention away from the fossil reptiles of the Lias to an overlying sequence of rocks, the Greensand, in search of fossil shells:

> 'You and Mr. De la B. between you have given me the green Sand Mania, both Miss Philpot & myself have been beating bit of green Sand to pieces to find shells, and after our return we sit down an turn over the leavs of Sowerby untill our poor heads is complete jumble of Bivalves, univalves &c &c.'

The 'Sowerby' she refers to is James and James De Carle Sowerby's *The Mineral Conchology of Great Britain*, a copy lent to her by De la Beche to help her identify her finds. The Greensand fossils occur in hard, rounded concretions called Cowstones which can be difficult to break open and the fossils are often not well preserved. Anning's letter continues 'I have not discovered anything since you left Lyme, as I have been to much taken up with green Sand'. This remark suggests that the letter dates from before December 1828 when Anning discovered the pterosaur fossil.

At this time, there was some considerable disagreement amongst geologists about the naming and order of the sequence of rocks containing the Greensand in southern England, a controversy which remained unresolved until 1836. Anning makes reference in her letter to some of the published scientific papers dealing with the Greensand controversy:

> 'I am greatly obliged to Mr. Murchison for his kind promise to lend me a copie of his memoir Mr. D.[e la Beche] has also promised me his and I hope by some means to get a peep at Dr. Fittons.'

Clearly she was aware of current research on the rocks with which she was familiar. That she was amused by the on-going geological arguments comes out in a comment in her letter, 'I do so enjoy an opposition among the big wigs'.[47]

In another letter, written to Charlotte Murchison before going to stay with her, Anning told her that she is sending her some fossils from the Greensand Cowstones but that her enthusiasm for these rocks had waned and that she had gone back to searching for fossil bones:

> 'Since I wrote to you I have not found much I soone lost the green sand mania, and returned to my old passion for bones, the other day having found a part of a plesiosaurus.'

She went on to describe the risks she faced when collecting this specimen, having nearly been caught by the rising tide:

> 'I was so intent on getting it out that I had like to have been drowned and the man I had employed to assist me, after we got home I asked the man why he had [not] cautioned me about the tide flowing so rapidly he said I was

ashamed to say I was frightened when you didn't regard it, I whish you could have seen us we looked like a couple of drowned rats, so woebegone it makes me cold to think of it.'

Being caught by the tide was an occupational hazard. Ten years later, in September 1839, Richard Owen recorded that he

'went to Lyme Regis, and there I met Buckland and Conybeare. They made me prisoner, and drove me off to Axminster, of which Conybeare is the rector. Next day we had a geological excursion with Mary Anning, and had like to have been swamped by the tide. We were cut off from rounding a point and had to scramble over the cliffs.'[48.]

Hers was a dangerous and exhausting occupation. In her letter of 25 February [1829] to Charlotte Murchison, Anning closes by apologising for not writing a better letter as, she says, 'I have walked ten miles to day I am so tired I can scarcely old the pen'.

In June 1829, De la Beche returned to Lyme after an eight-month tour on the continent, mostly spent in Italy. He had missed Anning's exciting discoveries of the pterosaur and a second complete plesiosaur, and the Geological Society meeting in February 1829 where Buckland gave his colourful paper about these finds and his description of fossil sepia and coprolites. De la Beche was amused by Buckland's fixation with coprolites and sometime around 1829 produced a small etching, *A Coprolitic Vision*, poking fun at Buckland who is depicted in academic gown and mortar-board, geological hammer in hand, gazing into a huge cavern which recedes into the distance, its roof supported by large, spiral coprolitic columns. To his right, a bear, a deer and hyaenas are producing copious coprolites-to-be, a reference to Buckland's recognition of hyaena coprolites in Kirkdale Cavern in Yorkshire in 1821. Pterosaurs fly through the cave above a lake or underground sea, relieving themselves as they go; a crocodile on a ledge drops its poop into the water, which is brimming with ichthyosaurs, no doubt defecating underwater. It is a combination of Anning's discoveries in Lyme and Buckland's in Yorkshire. [49.]

Despite the sale of the complete plesiosaur to the British Museum, the unreliability of the fossil business meant that by the end of the 1820s Anning was once more in financial difficulty. Help was, though, at hand, in the form of not only a new fossil discovery, but through support from her fellow geologists.

SEVEN

Mary and the Sea Dragons

A 'NEW BEAST'

ALTHOUGH THE ICHTHYOSAURS AND PLESIOSAURS are thought of as the jewels in Anning's fossil-finding crown, less well-known but equally remarkable was her discovery in December 1829 of a bizarre, new and scientifically undescribed – what she calls 'non-discript' – species of fossil fish. The specimen was about 17 inches (45 cm) in length, and seemed almost complete, lacking only its tail. It had large eye sockets, a scissor-like snout, claws or feelers, and wing-like fins resembling those of a ray. News of the discovery of 'an entirely new specimen of fossil organic remains' soon found its way into the newspapers as 1829 ended and 1830 began.[1]

Recognising its novelty, Anning wrote to her geological contacts to inform them of her find and with a view to finding a buyer. She described the specimen to her friend Charlotte Murchison, and wrote to Charles Lyell on 15 December 1829 to tell him of her discovery of a 'creature for a description of which I must refer you to Mrs Murchison, as it is impossible to describe it here.'

Anning must have written also to Buckland. At his suggestion she wrote on 30 January 1830 to J.S. Miller, the curator of the Bristol Institution, who had visited Lyme in 1824 and had bought a large collection of fossils from her, and later an ichthyosaur:

> '... by Dr Buckland's advice I offer my new specimen to Bristol Institution, I have disected a Ray since I received your letter, and I do not think it the same genus, the Vertebrae alone would constitute a different genus being so unlike any fish vertebrae ...'[2]

Anning's knowledge of the Lias fossils and their anatomy was such that she recognised her find as new species of fish, in fact a new genus, even going so far as to dissect a modern ray for comparison. This is Anning the scientist at work.

Another letter to Buckland urged him to visit Lyme to see the specimen for himself, telling him that she had just sent a drawing of it to Cuvier – though as with her recent plesiosaur she felt that so unique a specimen should stay in England.[3.]

In due course Buckland went to Dorset, followed by a cloud of rumours that do little to dispel the petty jealousies circulating amongst his fellow geologists. Here is Charles Lyell writing to Gideon Mantell in April 1830 with news of Buckland's visit to Lyme:

> 'so there is something in the wind – a paper on the new beast perhaps, that fish-like concern which Mary Anning wants to make a grand wonder of, and the Dr a memoir, I suppose.'[4.]

Lyell's sarcasm suggests that Anning is making a 'grand wonder' of her specimen in order to sell it for as high a price as possible and that Buckland intends to write a scientific paper about it. As it happens, there is no indication that Buckland ever did so. His friendship with Anning was long-standing, their interest in the Lias fossils mutual, and in recent years they had worked together on fossil sepia, coprolites and the pterosaur. Never once, in all his papers and letters, is there even a whisper of Buckland taking advantage of Anning to add to his academic reputation. Anning had known him since childhood, and he was usually the first person she notified after finding a new or significant fossil. Their relationship was more symbiotic, with Anning making the discoveries and Buckland bringing them and her to the attention of the geological community, finding buyers for her and acting as a go-between, viewing specimens on behalf of potential buyers like the Duke of Buckingham. In his papers he generally acknowledged Anning as the finder.

The Bristol Institution was interested in acquiring the new specimen and the Reverend David Williams went to Lyme on their behalf to examine it, offering £35 which Anning declined. Negotiations seem to have continued through 1830 and into early 1831, the delay persuading Anning to send a sketch of the fossil to Adam Sedgwick in Cambridge,

> 'Sir I do not know if you have heard of a new fossil now in my possession. It is a skeleton with a head like a pair of scissors Vertebrae like an ecrinite thin as a thread of which there are two 100 & 52 and the tail wanting the greater portions of six claws or felers and winged like fins sternum simple composed but of two bones also the pelvis the vertebrae skin and snout covered with tubercles like those of the ray tribe which it strongly resembles in some parts and wholly differs in others the teeth are like the tubercles on the body except that they are larger and crooked it is quite unique analogous to nothing yet approaching to fishes insects birds and animals about a foot and half in

length of which the underneath scratch is a faint resemblance, and being the only one in Europe price 50£.'

The 'scratch' was her sketch. Anning was capable of producing very competent drawings as a means of encouraging sales. She also circulated a handwritten flier describing the specimen – unpriced – and two ichthyosaurs she had for sale.[5]

Anning's 'non-discript' fish did eventually make its way to Bristol. It was purchased for the Bristol Institution by a wealthy Bristolian property-owner, John Naish Sanders, for an unknown sum, but mostly likely somewhere around £40. Sanders had been involved with the Institution since its beginnings and was, perhaps, its most generous donor and benefactor.[6]

The specimen was first described by Bristol physician and Institution member Dr Henry Riley in a paper read to the Geological Society of London in May 1833. Crediting Anning with the discovery, but placing it in 1831, Riley's identification of the fossil was largely based on Anning's ideas of the affinities of the specimen, deciding that it was a cartilaginous fish, despite other suggestions that it was a reptile or even a bird. It had similarities with sharks (squales) and with rays, but differed from both so Riley named it *Squaloraia*. Others disagreed with his identification and it was not until the 1870s that it was recognised as belonging to a group of cartilaginous fish sometimes called ghost sharks. In 1843 Louis Agassiz named it *Squaloraja polyspondyla*.[7]

The specimen purchased by Sanders was missing its tail and this Anning had promised to send to him should it be found. Eventually he tracked it down to the collection of the Philpot sisters in Lyme and wrote to them to ask if it might be reunited with rest of the specimen in Bristol:

'I believe that you are already aware that I purchased of Miss Anning, some years ago, a rare fossil, described by Monr. Agassiz, and of which there is a large plate in the last report of the Geological Transactions. If Miss A. Had not forgotten her promise to me, a portion of the tail, subsequently found, would now be attached to the fossil. Should you be of the opinion that science would be promoted by a reunion, I should be most happy to pay any expense that you have incurred, for the fragment in your possession.'

The Philpot sisters were not persuaded, which is just as well in light of subsequent events. The tail went with the rest of the Philpot collection in 1880 to Oxford where it now resides in the University Museum of Natural History. Like much of Bristol city centre, the rest of *Squaloraja* fell victim to German bombs on the night of 24-25 November 1940.[8]

DURIA ANTIQUIOR

It took over a year for Anning to sell her fossil fish and little remained of the 100 guineas she had been paid for her plesiosaur. Once again, she was running out of money. Aware of her predicament, and that she was too proud to ask for help, William Buckland and Henry De la Beche came up with a novel way to raise money for her and to pay tribute to her discoveries. De la Beche produced another drawing, a watercolour, inspired by Buckland's description of life in the Lias of Lyme in the distant past and the fossils discovered by Anning. They called this image *Duria antiquior*, 'an earlier or more ancient Dorsetshire'.

Coiled-shelled ammonites float on the sea surface and pterosaurs soar through the air, although one unfortunate animal is being snatched from the air by a plesiosaur. In the centre of the scene, a plesiosaur's neck is being broken by the jaws of a large ichthyosaur while, under water, another ichthyosaur devours a fish. Three or four different species of ichthyosaur are illustrated, reflecting De la Beche's work on these animals a decade earlier. Faeces are being produced from the nether regions of several of the ichthyosaurs and the central plesiosaur and settling on a sea floor littered with ichthyosaur skeletons and shells, all destined to be fossils of the future. All of the main elements in this scene represent Mary Anning's discoveries in the rocks of Lyme Regis brought to life through De la Beche's pen and brush and Buckland's imagination.[9]

De la Beche arranged for his watercolour to be reproduced as a lithograph; the image was transferred to stone by the artist George Scharf and printed by Charles Hullmandel, and prints were sold for £2 10s, so it was an expensive purchase. Buckland sold impressions to fellow members of the Geological Society, and the proceeds went to Mary Anning. In a letter to De la Beche on 25 May, probably in 1830, he writes 'I have sent M. Anning £5 for copies sold to Stokes, Lonsdale and Broderip'. The print was well-received: on 13 May 1830 Charles Lyell wrote to Gideon Mantell

> 'Have you seen the admirable plate of Buckland & Delabeche's ... Duria antiquior or Dorsetshire as it existed at the era of the lias? It is really a glorious restoration & has done much to popularize the subject ... The Plate was done for Mary Anning & will put much money in her pocket.'[10]

Although it was generally accepted that *Duria* was being sold to benefit Anning, the lithograph carries no inscription or dedication to that effect. Early prints merely have the title while later ones also have a numbered key identifying some of the animals.

Buckland was delighted with the picture and wrote to De la Beche to tell him how his university teaching was much improved by it. He used an enlarged version, perhaps drawn by his wife, as a visual aid in his lectures at Oxford and had additional small copies run off so that he could distribute them after his classes to remind his students of what his lecture had been about, an early example of a lecture handout. Buckland also sent a hundred copies of the lithograph to Anning in Lyme Regis, presumably so that she could sell them in her Broad Street shop.[11]

We have no record of how many copies of *Duria antiquior* were printed and sold, or how much money Anning received. Nor do we know what she thought of it, though one can easily imagine her delight at seeing her fossils featured as they were in life. The cultural impact, though, of *Duria* was far-reaching. It was distributed, copied and plagiarised widely and it and similar reconstructions became a common feature of popular books on fossils through the nineteenth century. Its influence is still with us today in the form of the computer-generated imagery used in television and film to reconstruct dinosaurs and life in the geological past.[12]

Buckland returned to Lyme with his family in the summer of 1830, scouring the shore in a fruitless search for fossils. 'I have been here a month with my wife and children,' he wrote to De la Beche, 'but have found no great novelty of the subterranean world.'[13]

The precariousness of the fossil trade is underlined in a letter of October 1833 from Anning to Charlotte Murchison. Mrs Murchison had sent an order for some specimens, but prior to the winter storms, Anning had little material in her shop. Anning replied to her,

'I feel gratefully obliged by the order of specimens for your friend but at present I have not anything fine or rare, for the last year I have been very unsuccessful but hope as the stormy season is coming on I may be so, may I ... ask for a list of the specimens you most wish yr friend to have'.[14]

The winter months of 1830-31 lived up to expectations. In December she discovered another near-complete plesiosaur skeleton, clearly a different species from those she had previously found. Four days before Christmas she told Buckland of her new find:

'I write to inform you that in the last week I discovered a young Plesiosaurus about half the size of the one the Duke had, it is without exception the most Beautiful fossil I have ever seen. The tail and one paddle is wanting (which I hope to get at the first rough sea) every bone is in place, in short if it had been made of wax it could not be more beautiful, but I should remark that the head is twice as large in proportion as those I have hitherto found. The neck has a most graceful curve and what makes it more interesting is that resting on the bones of the pelvis is, its Coprolite finely illustrated.'[15.]

The presence of a coprolite within the pelvic region made this a remarkable find. Anning's enthusiasm comes through in her letter, appreciating the beauty of the specimen, almost as a sculpture, a wax model, with its curved neck, as do her powers of analysis and her expertise, recognising that it is not only a new species but also a juvenile.

Anning must also have informed Charlotte Murchison of the discovery who in turn may have acted as Anning's unofficial agent in the search for a purchaser. On 8 February 1831, Anning received a letter from Charlotte to say that Adam Sedgwick in Cambridge was willing to make an offer. He was too late. The plesiosaur had been sold within a month of its discovery. Anning wrote to Sedgwick to apologise:

> 'I am exceedingly sorry Mrs M. did not get my letter which I sent by a Neighbour on the first of this month, stating that Lord Cole had become the purchaser of the plesiosaurus he did not say wether it was for himself or Mr Egerton.'[16.]

Cole had reportedly paid Anning £200 for the specimen.

Lord Cole was Viscount Cole, William Willoughby Cole, later the 3rd Earl of Enniskillen. He and Philip de Malpas Grey Egerton met when both were students at Christ Church, Oxford, and attended Buckland's geology classes. The pair developed a lifelong friendship and in the summer of 1830 set out on a tour of the continent, encouraged by Buckland, to study cave deposits. A meeting with the Swiss geologist and fossil fish expert Louis Agassiz in Munich redirected their interest from caves to fossil fish, which they took to with enthusiasm bordering on obsession. They collected jointly, sometimes sharing parts of the same specimen or tossing a coin to decide into whose collection it would go. While their main focus was fossil fish, they did occasionally acquire others, such as this new plesiosaur.[17.]

The specimen was illustrated by William Buckland in his 1836 book, *Geology and mineralogy considered with reference to natural theology*: 'a beautiful specimen of Plesiosaurus macrocephalus hitherto undescribed, found in the Lias marl at Lyme Regis by Miss Anning, and now in the collection of Lord Cole.' Buckland did not give a full description of the new plesiosaur, but merely listed its differences from *Plesiosaurus dolichodeirus*. The larger and longer head led Buckland to give it the species name '*macrocephalus*' ('large-headed').[18.]

Also in Cole's collection was a lobster-like fossil crustacean bought from

Anning in the early 1830s. It was described in a paper to the Geological Society in 1835 by William John Broderip, a Bristol-born barrister and naturalist. He recognised the fossil as new to science and named it *Coleia*, a new genus, after Lord Cole who now owned the specimen, even though he records that it was 'found in the lias at Lyme Regis by Miss Anning'. In the same paper Broderip also describes a new species of starfish, a brittle star, which had been found at Bridport about 8 miles east of Lyme and which he named *Ophiura egertoni* after Cole's collecting partner. Broderip also examined three specimens belonging to the same species bought from Anning by Adam Sedgwick. While we usually associate Anning with her discoveries of vertebrate fossils, we should remember that she also made significant finds of invertebrates like these crustaceans and starfish, as well as of crinoids.[19.]

The main fossil collections of both Cole and Egerton were eventually acquired by the British Museum following the transfer of its natural history departments into the new purpose-built Natural History Museum in South Kensington, which opened in 1881. Egerton's was purchased from his estate in 1882 following his death, and Cole's collection of nearly ten thousand specimens, including *Plesiosaurus macrocephalus,* was purchased for £3,500 in 1883. The Anning-Cole plesiosaur is a juvenile – as recognised by Anning herself – and is about half the size of the largest known example of this species.

In her letter to Sedgwick explaining the sale of the plesiosaur to Lord Cole, Anning described other specimens she had available, including an *Ichthyosaurus* skull 'about 15 inches in length with six cervical vertebrae attached set in a box and very good specimen price two pound ten'; a headless ichthyosaur skeleton, 'Vertebrae paddles Sternum and ribs in good order and on the bones of the pelvis is the Coprolite price 3£.' In this letter she also describes, with a sketch, the new fossil she had found which later went to the Bristol Institution and was named *Squaloraja polyspondyla* by Louis Agassiz. Sedgwick purchased the two ichthyosaur specimens, as recorded by his rough account books: 'Mary Anning for fossils – £5. 10. 0.'[20.]

Sometime between February and May 1831, Anning found a particularly fine ichthyosaur. Buckland again seems to have been amongst the first to know, informing De la Beche in a letter of 1 May:

'Mary Anning has found another small & very perfect Ichthyosaurus for wh. she asks £40 – the best she says she has ever seen. I have sent notice to Sedgwick, but she has promised the first offer to Mr Botfield, & Mr Saul has also applied for the choice of the next.'

There remained a demand for ichthyosaur specimens. Although their novelty had worn off, prices had fallen and Anning had a waiting list for her finds. Northamptonshire collector Beriah Botfield had first refusal on this specimen, whilst next on the list was William Devonshire Saull, a London wine merchant who built up his own museum of fossils and antiquities, some twenty thousand specimens in total.[21]

In the autumn of 1831, Anning sold an ichthyosaur to the eminent surgeon and comparative anatomist Sir Astley Cooper who called at her shop on his way home from a tour of the West Country. Cooper had Anning pack the specimen and send it to his home in Hemel Hempstead. Anxious to hear of its safe arrival, she wrote to him, probably sometime in late 1831:

> 'Mary Anning's best respects to Sir Astley Cooper would feel greatly obliged by a line to say wether the Skeleton arrived safe'

A 'small *Ichthyosaurus intermedius*', it may be the specimen mentioned by Buckland to De la Beche. Charles Lyell referred to it in a letter to Gideon Mantell in April 1832: 'I met Sir A. Cooper ... at dinner yesty. who talked of Mary Anning & of an icthyo. he had bought.' Cooper was much taken with Anning, whom he described as 'a great genius, whose knowledge, zeal, and ability are really surprising.' The specimen he purchased was later illustrated in Buckland's 1836 *Bridgewater Treatise*, and has recently been identified in the collection of the Natural History Museum in London.[22]

OUT OF LYME

Mary Anning's 1829 visit to London was, she says, her first time out of 'the smoke of Lyme' and we have a record of only a handful of other occasions when she left the town. At some point, Anning may have travelled to Watchet on the Somerset coast. A specimen of a fossil squid ink sac, illustrated by William Buckland in his 1836 *Bridgewater Treatise,* is captioned 'found by Miss Anning in the Lias near Watchet, Somerset'. The specimen, which still exists in the collections of Oxford University Museum of Natural History, is catalogued as having been collected at Watchet in 1828 by Anning. However, as that precedes her London visit the date may be incorrect. Another possibility is that Anning had not been to Watchet at all, but had been given the specimen and Buckland mistakenly thought that she had collected it herself.[23]

While almost all of Anning's fossil collecting was undertaken close to Lyme, occasionally she ventured further afield, to Dowlands and Culverhole Point in Devon to the west and to Golden Cap or even almost as far as Bridport to the east. In the autumn of 1840 Louis Agassiz identified some fossil fish which Anning had collected from Triassic rocks which lie below the Jurassic Lias limestones. These probably came from Culverhole Point, about four miles west of Lyme. We know from a letter from Elizabeth Philpot to William Buckland that Anning's expeditions took her east along the coast. In June 1835, probably at Thorncombe Beacon or Eype Mouth she found some crinoid fossils which she recognised as a new species. While it is perfectly possible that the fit and strong Anning walked there and back in a day, it is more likely, given the distance, terrain and tides, that she travelled by coach or cart, or perhaps even by boat, assistance provided by friends in Lyme.[24.]

Another Anning excursion from Lyme is recorded in an intriguing entry in the diary of Lady Margaret Paterson, wife of Major-General Sir William Paterson. On Thursday 16 June 1831, Lady Paterson records that her daughter, Margaret, went with her father and a friend, Miss Kekewich,

> 'to Miss Barings to tea to meet some friends who had dined, amongst whom were Lady Duckworth, Major Read etc, and a Miss Anning, a great geologist, who has made some remarkable discoveries.'

Susannah-Catherine, Lady Duckworth of Wear House, Topsham, was the widow of Admiral Sir John Thomas Duckworth and daughter of William Buller, Bishop of Exeter, while their host, Miss Baring was a member of a wealthy banking family which included several members of parliament. Exeter is a good 30 miles (50 km) from Lyme. How did Mary Anning come to be invited to dine with fellow guests who came from several prominent Exeter families? Whatever the reasons, it certainly confirms her celebrity, and demonstrates her self-confidence in accepting the invitation. The fact that Lady Paterson refers to her as 'Miss Anning' in the same manner in which she refers to 'Miss Baring', rather than simply 'Mary Anning' as she might usually address the lower classes, implies both her social acceptance by the middle and upper classes and, at least to a small degree, some upward mobility on Anning's part.[25.]

We also have a hint of another visit out of Lyme made the following year. On 5 June 1832, William Buckland wrote to Joseph Anning to arrange to have some specimens sent to him in Oxford. An ichthyosaur, he suggests, can be sent by wagon if Joseph feels confident that it will not be damaged by the journey, but he writes, 'the Sepias and Coprolites you have for me had better wait to be packed & sent when your sister comes Home. Let them come by Coach & desire her to be most careful in packing them'. He concludes by sending his regards to Joseph's mother, Molly. The question arises: where was his sister? She was clearly away for several days, and Buckland was aware that she was not at home. This letter also illustrates that despite his work as an upholsterer, Joseph was still helping out with the family fossil business in the 1830s.[26.]

Mary Anning was certainly at home by late June when the Sussex doctor and fossil collector, Gideon Mantell visited Lyme in June 1832. It was a beautiful summer's day when Mantell, travelling with a friend, probably the Reverend David Williams, Rector of Bleadon near Bristol, 'descended by a fearfully precipitous loose and dangerous road to Lyme Regis' after a long and arduous ten hour journey, having left Bristol early that morning. They arrived at their inn at six p.m., and

> 'after taking refreshment and washing, we sallied out in quest of Mary Anning, the geological Lioness of the place. We found her in a little dirty shop, with hundreds of specimens piled around and in the greatest disorder. She, the presiding Deity, a prim, pedantic vinegar looking, thin female, shrewd, and rather satirical in her conversation.'[27]

Such is Anning's reputation that Mantell regards her as a 'geological Lioness', but his journal clearly shows his disappointment at their first encounter. Mantell's impressions are largely at odds with other descriptions of her. His view of her as 'prim, pedantic, vinegar looking' may exhibit Mantell's own casual prejudices, employing epithets that tended to be applied to spinsters. Perhaps Anning displeased Mantell by correcting him on some matter, leading to the description of her as 'pedantic'. Later accounts of Anning during this period suggest she had a firm belief in her own opinions, certainly where fossils were concerned, and did not hesitate to speak her mind. It has been suggested that Mantell may have mistaken Molly Anning, Mary senior, for her daughter but given the difference in age, that does seem unlikely.[28]

In midsummer, long after the winter storms, her shop often lacked much that was special. Mantell's diary continues,

> 'She had no good specimens by her; but I purchased a few of the usual Lias fossils. Having heard a fisherwoman had some fossils we sought her out: she had a portion of the vertebral column of a Plesiosaurus dolichodeirus, and other bones which I bought....'

Early the following morning, before departing by boat for Portland, Mantell went out onto the foreshore to examine the Lias cliffs which he 'had heard, read and known so much', but never before seen, and found ammonites and brachiopods, but no bones.

HAWKINS HITS LYME

In July 1832, Anning welcomed perhaps the most eccentric of all the customers to enter her shop, Thomas Hawkins. Anning historian W.D. Lang called

Hawkins, 'a collector of great energy and eccentricity, a man who did nothing by halves, who may be said to have moved in an atmosphere of superlatives'. He was very deaf and delusional – he claimed to be the 'Rightful Earl of Kent'; he was provocative and quarrelsome, provoking feuds and even minor riots; litigious; and overtly sycophantic in his attitude to the scientific and academic establishment. Gideon Mantell described Hawkins as 'very romantic, very weak, very good natured, & I fear very headstrong'.[29.]

Thomas Hawkins was born at Glastonbury in Somerset in 1810, the son of a farmer and cattle dealer. His father died in 1830 and the following year, Hawkins entered Guy's Hospital in London as a surgeon's pupil. Two years later he was living at Sharpham Park, the former home of the author of *Tom Jones*, Henry Fielding, on the Somerset Levels near Street. Advised to try sea-bathing to alleviate his deafness, Hawkins travelled to Charmouth. He later wrote that while there 'there were big storms, rock ledges were torn up and fossil dragons exposed on the beach'. Nearer home, quarries around Street worked the same Lias limestones and 'dragons' were being found there too. Hawkins had collected fossils as a child, but now began to concentrate on the marine reptile fossils from the Street quarries, making the acquaintance of the quarrymen and paying them to set aside any that they might find. Very quickly, Hawkins built up an impressive collection.

In 1832 Hawkins was elected a Fellow of the Geological Society, and in the summer, went to Lyme to acquire more specimens for his collection. He arrived just as Mary Anning was excavating part of the skull of an ichthyosaur from Church Cliffs. Hawkins determined to extract the remainder of the skeleton for himself. He purchased the part Anning had already removed and with permission of the landowner, James Edwards senior, employed some of the quarrymen who worked the foreshore limestone quarries and set about a large-scale excavation involving construction of a track down the cliff.[30.]

He describes the event in a book he published in 1834, *Memoirs of Ichthyosauri and Plesiosauri, extinct monsters of the ancient Earth*, in his own inimitable style:

'The sun rose bright on the 26th day of July, 32, and the morning mists were hardly rolled from the hill's side ere many men busily engage with spade and pick-axe to humble the doomed summit of this cliff. Progress was also made on the following day, when people from the adjacent country flocked to witness the execution of a purpose which seemed to stagger their faith in our rationality. By next day's noon twenty thousand loads of earth, cast from the crown of the rock, constitute a good roadway to the beach from that part of it to which we had dug, and a few minutes more suffice to demonstrate the wonderful remain I tell of. Who can describe my transport at the sight of the colossus! *My* eyes the first which beheld it! who shall ever see them lit up with the same unmitigated and enthusiasm again! And I verily believe that the uncultivated bosoms of the working-men were seized with the same contagious feeling, for they and the surrounding spectators waved their hats to a hurra, that made hill and mossy dale echoing ring'.

Hawkins' enthusiasm and inexperience – and his wealth – meant that he was prepared to take on a highly ambitious and labour intensive excavation of an ichthyosaur that the more experienced, pragmatic (and impoverished) Anning had clearly thought not worth pursuing. During the excavation the bones and the marl in which they were embedded broke into many small fragments and it was only with Anning's help that Hawkins was able to complete their packing before nightfall. The fossil in its rock matrix weighed a ton.

Hawkins had the specimen transported to Sharpham Park where it arrived at 6 am on 1 August and immediately he began to unpack it and place the several hundred pieces in order. He then spent several weeks preparing the specimen, removing the rock matrix to expose the bones. Once this was done, the slabs of rock with the bones were embedded in plaster within a wooden frame, the whole thing weighing half a ton. At over 20 feet (6 m) in length, it is one of the largest and most complete Lias ichthyosaurs known, and today it is on display in the Fossil Reptile Gallery of the Natural History Museum in London. Like the Anning's first ichthyosaur, it belongs to the species named *Ichthyosaurus platyodon* by De la Beche and Conybeare, although its size and tooth-shape led this species to be put into a new genus called *Temnodontosaurus* ('cutting-tooth lizard') in 1889. *Temnodontosaurus* was probably an open ocean species and the largest and most ferocious predator of the Lias seas, reaching lengths of over 30 feet (9 m). Its prey included other ichthyosaurs.[31.]

Roderick Murchison commented on this specimen in his retiring Presidential Address to the Geological Society in February 1833:

'A recent discovery of Miss Mary Anning, that indefatigable purveyor to the store-houses of our science, has furnished Mr T. Hawkins with the disjointed fragments of an animal, which upon being reintegrated, proves to be the largest individual of the *Ichthyosaurus platyodon* ever yet found entire upon our shore.' [32.]

Murchison's 1833 lecture must have been impressive; Anning heard about it from one of the Society's Fellows, William Hutton, and was keen to see the published version. In a letter to Charlotte Murchison on 11 October 1833, Anning asks for a copy of 'Mr. Murchison's last anniversary speech, I long to see it for Mr. Hutton told me it was the best he had ever heared and that Mr.

Murchison looked like a God when he made it, which I most cordially believe for Mr. Murchison is certainly the handsomest piece of flesh and blood I ever saw.' A curious remark to make to a man's wife. But the forthright Anning was obviously comfortable enough in her friendship with Charlotte to comment on her husband's good looks.[33.]

Hawkins returned to Dorset in the wet summer of 1833, and called upon a local fossil collector:

> "Have ye sid my animal sir," said the fossilist Jonas Wishcombe of Charmouth as I called at his house to enquire if he had anything worth buying; – "I should like vor yer honour to see 'im." My heart leaped to my lips- "animal! animal! where?" "Can't be sid to day sir – the tide is in" "What nonsense! I must instantly – come, come along." "Can't see 'un now yer honour – the tide's rolling atop'o'im fifty feet high." "In marl or stone?" "Why in beautiful ma-arl". "Washed to death" – and I threw myself in despair upon a chair. How often have I reflected upon the very Bedlam impetuosity of my passions at that moment: – the chaffed sea rolling over an Ichthyosaurus and remorselessly tearing it to a thousand atoms – a superb skeleton of untold value titurated to sand by a million pebbles, such was the Promethean idea-torture of my rebel imagination'.

At low tide the following day, accompanied by Mary Anning, Hawkins walked from Lyme to Charmouth so that Wishcombe could show him where the fossil lay, its backbone just visible in the rock. Wishcombe had known about it for months but considered that it was not worth trying to extract. Hawkins wrote that Wishcombe

> 'chuckled when I gave him a guinea earnest-money, convinced that he had made brave of a discovery that no one could render useful ... "You will never get that animal," said Miss Anning, as we made our devious way towards Lyme through the mist and flashing spray, "Or if you do, perchance, it cannot be saved." My eyes glance upon the intellectual countenance before me – the words of those lips were I knew oracular as those of a Pythoness and my heart fainted within me. She saw my change of blood and stopped – "because the marl, full of pyrites, falls to pieces as soon as dry." I revive "That I can prevent," "Can you?"'

Again we have the experienced Anning advising the novice Hawkins, here about the fragility of the fossils preserved in the marl, how they crumble when dry and how the pyrite – iron sulphide or 'fool's gold' which gives some ammonites their golden sheen – in the rock can cause it to break up. Such was Hawkins impetuosity that, although he recognised her advice as having the authority of an oracle, he disregarded it nonetheless. The contrast of his

THE FOSSIL WOMAN

description of Anning's 'intellectual countenance' with Mantell's 'prim' and 'vinegar-looking' is striking.

In his book, Hawkins decided that the *Ichthyosaurus* species names which had been proposed by Conybeare and De la Beche and were by then in common usage, were inadequate. Based on the shape of the limb bones, Hawkins took it upon himself to invent new species names: *chiroligostinus*, *chiropolyostinus*, *chirostrongulostinus* and *chiroparamekostinus*. These clumsy and unnecessary names never caught on. Geologists like Conybeare, De la Beche and Buckland, all of whom were subscribers to Hawkins' book did not take him seriously. In a postscript to a letter to Buckland, Conybeare wrote, 'What capital fun Hawkins book is. I only wish it had been published before Walter Scot died. It might have furnished him a new character, a Geological bore far more absurd than all his other ones put together.' Clearly Conybeare was not a fan of either author. Gideon Mantell called the book 'strange but splendid'.[34.]

Hawkins published another volume in 1840, *The book of the great sea dragons, Ichthyosauri and Plesiosauri, gedolim taninim, of Moses. Extinct monsters of the ancient earth.* In it Hawkins further revised his naming of the ichthyosaurs, dropping *Ichthyosaurus* as a genus and replacing it with *Oligostinus*, so that his large specimen from 1832 was now called *Oligostinus chiroligostinus* – but only by him. If his 1834 *Memoirs of Ichthyosauri and Plesiosauri* were considered eccentric, *The book of the great sea dragons* is downright odd, a strange mix of religious ramblings and palaeontology. One late nineteenth century writer on geology described Hawkins' two volumes as 'a curious mixture of bigotry, conceit, and unrestrained fancy'. More recently, the American evolutionary biologist the late Stephen Jay Gould suggested that the two books 'display a mind in disaggregation' and that while the 1834 book is 'florid enough, but still tractable' the 1840 book is 'all but unreadable'.[35.]

The illustrations, however, are wonderful, with large lithographed plates of his specimens, many by the renowned London lithographic artist, George Scharf. Each book has a frontispiece, a reconstruction of a Lias scene of plesiosaurs and ichthyosaurs. That in the 1834 *Memoirs* is a placid and tranquil view by John Samuelson Templeton while the plate in the 1840 *Great sea dragons* is anything but. In this dark, dramatic and demonic scene, a typically gothic and nightmarish imagining by the artist John Martin, an ichthyosaur is set upon by two wild and maniacal plesiosaurs while a group of pterosaurs feasts on the carcass of another ichthyosaur stranded upon a shoreline. The difference between the two images is stark, and one is compelled to wonder how much this change is indicative of Hawkins' state of mind, or of Martin's.[36.]

Hawkins' books each relate to the purchase of his specimens in two lots by the British Museum, one in 1834 and the other in 1840, for a combined sum of £3,115. 5s. In 1833 Hawkins had offered his first collection to the museum for £4,000, but the trustees declined. Undaunted, Hawkins began lobbying anyone who might influence their decision and was eventually successful, although he did not receive the amount he had sought. William Buckland, along with Gideon Mantell, had been asked to value the collection, which he estimated to be worth £1,310.

Several of Hawkins' specimens, including some coprolites now in the Natural History Museum, were collected by and purchased from Anning. Hawkins was generous and honest in his acknowledgement of the help he received from her – albeit in flowery prose – and of the debt owed to her by geologists like Buckland, Conybeare and De la Beche:

> 'But although many obligations are owing to the zealous efforts of these justly eminent personages, yet it must never be forgotten how much the exertions of Miss Anning of Lyme, contributed to assist them. This lady, devoting herself to Science, explored the frowning and precipitous cliffs there, when the furious spring-tide conspired with the howling tempest to overthrow them, and rescued from the gaping ocean, sometimes at the peril of her life, the few specimens which originated all the fact and ingenious theories of those persons, whose names must ever be remembered with sentiments of the liveliest gratitude.'[37.]

Soon after the purchase of Hawkins' collection in 1834, questions were raised about some of the specimens. How much was actually fossil bone, and how much was plaster reconstruction by Hawkins? Gideon Mantell had noted the extent of restoration on Hawkins' large ichthyosaur when he saw it in London in December 1832:

> 'Went to Mr Hawkins and saw his splendid specimens – one of them an Ichthyosaurus platyodon 25 feet long: I must however observe that this gentleman Mr H. restores them too much: he makes nothing of putting on an arm or a tail, or a rib, where they may be wanting: he does not do this without authority; yet still I think it is objectionable when art is allowed to interfere so far.'[38.]

Mary Anning, too, was concerned about the authenticity of Hawkins' restoration work. In a letter to Charlotte Murchison of October 1833, she wrote,
'if Mr. Hawkins has set the last specimen of Icht platyodon as it lay in the

cliff it will be a most magnificent specimen, but he is such an enthusiast that he makes things as he imagines the ought to be; and not as they are really found, the platyodon that I in part gave him was to large for my poverty & I would not have trusted to his making up, though very much broken it might be made a splendid thing without any addition.'[39.]

Once the collection had arrived at the museum, Charles Konig discovered the extent of Hawkins' plaster restoration on his large ichthyosaur and notified the trustees of his 'discovery of rather a vexatious nature' and that in his opinion the specimen was 'altogether unfit to be exhibited to the public without derogation from the character of the British Museum'. The story leaked to the press and was the subject of much scrutiny by the House of Commons Select Committee on the British Museum in July 1835. Mary Anning's fears about Hawkins' overuse of plaster were justified.[40.]

ANNA MARIA PINNEY

From late 1831 Mary Anning became friends with a young woman visitor to Lyme, Anna Maria Pinney, whose diary gives us a rare insight into Anning's thoughts and attitudes. Pinney had moved to Lyme from Somerton Erleigh in Somerset in October 1831 to canvass support for her older brother, William Pinney, who was standing as the Whig candidate for election as Lyme's Member of Parliament at the 1832 General Election, the first after the Reform Act. Anna Pinney kept a journal in which she recorded her activities, who she met, what they talked about, and snippets of local gossip. In her conversations with Pinney, Anning seems to have been candid and trusting, sharing confidences but perhaps not expecting her remarks to be recorded. [41.]

On Tuesday 25 October 1831, Pinney noted in her journal that she 'went out at 11 o'clock fossilizing with Mary Anning ... a woman of low birth, her brother is an upholsterer here and one of William's voters.' She then gives an account of how Anning had been struck by lightning as a baby and how, after her father's death, she had been given half a crown for a fossil she had found on the seashore. These are familiar stories, the same as those told by Anning to Frances Bell in 1824, and ones it seems she recounted to any new friends eager to learn something of her humble background.

Pinney's journal goes on to paint a picture of Anning as a complex figure: proud, confident, opinionated, intelligent, independent, pious, kind and generous, but equally appearing conceited, resentful, gossiping, blunt and indiscreet, unafraid of causing offence and now to an extent at odds with her social peers:

'She has been noticed by all the cleverest men in England who have her to stay at their houses, correspond with her in geology &c. This has completely turned her head, and she has the proudest and most unyielding spirit I have ever met with. Much "learning has made her mad". She glories in being afraid of no one, and in saying everything she pleases. She would offend all the world, were she not considered a privileged person.'[42.]

In this journal entry, more than anywhere else, we get a sense of Mary's intellectual confidence, independence and sense of her own self-worth; 'unyielding' perhaps in that she was unwilling to concede in debate, regardless of the identity of her opponent. A more submissive attitude would have been expected of her – both as a woman and a member of the working class. Perhaps she appeared combative and arrogant, or perhaps she simply confounded Pinney's expectations of how a woman should behave. This suggests that Anning's celebrity gave her licence to flout convention, perhaps beyond what was normally socially acceptable, although her standard of what was considered acceptable might have been different from that of a young middle class woman, giving rise to misunderstanding. Anning was now in her early thirties, and by far the most famous person living in Lyme. Her reputation was secure, and she had the confidence to say what she thought.

George Roberts devoted over six pages to her in his 1834 *History and Antiquities of Lyme Regis*, concluding 'by mentioning Miss Anning has a fossil *depôt*, where all the specimens are sold at a very reasonable rate'. He praises her 'genius for discovering where the Ichthyosauri lie embedded' but then makes a curious remark:

'Though the subject of this notice [Anning] would not be disparaged by a description more strictly personal, yet such might be unpleasant.'[43.]

We can only guess to what Roberts alludes; there is something he is not telling us about her, perhaps the traits noted by Pinney, or some disagreeable side to her character and personality. He may have been reluctant to say more. After all, he had done much to put her on the map. She was a friend, neighbour, and subscriber to his book.

In an echo of a comment Anning made to Frances Bell in 1824 that 'the world has used her so unkindly,' Pinney records that

'She says the world has used her ill and she does not care for it, according to her account these men of learning have sucked her brains, and made a great deal by publishing works, of which she furnished the contents, while she derived none of the advantages.'[44.]

Here, finally, in the company of her nineteen-year-old friend, Mary Anning can allow her grievances to surface. Her fellow scientists had private means, she didn't. The years spent scrambling along the cliffs, hammer and collecting basket in hand, her dog Tray at her feet whatever the winter weather, had, she felt, enhanced their reputations, with the prospect of honours, appointments, positions of power. Only in one regard is she is wrong. None of those who published papers on her specimens gained much financially, as publishing in scientific journals was not remunerative.

Certainly Anning's discoveries had made her name in geological circles. She was not directly credited with her finds in the publications of only one or two geologists, notably Conybeare, but this seems not to have affected their friendship. Others like Buckland, De la Beche, Egerton, Broderip, Owen, Riley and Hawkins all acknowledged Anning as the source of specimens they describe in their papers and books, and it was common knowledge at the Geological Society and amongst the fossil collectors of the west of England that she had provided almost all of the new discoveries. As Edward Pidgeon pointed out in his 1830 book, *The Fossil Remains of the Animal Kingdom*, 'it is to her almost exclusively that our scientific countrymen ... owe their materials on which their labours and their fame are grounded, nor ... will they be unwilling to admit that they are indebted for some portion of their merited reputation to the labours of Mary Anning.'[45.]

But an acknowledgement in a scientific paper brings few material benefits. The lack of money was a constant niggle. Her fame brought visitors to Lyme – tourists wanting to rub shoulders with the legend that was Mary Anning, geologists looking to purchase specimens or go out fossilising with her. It did make her a celebrity, a source of interest and a recognised authority, but how much was she compensated for her expertise? It is hardly surprising that she felt she was being exploited.

Pinney also noted that Anning

> 'says she stands still, and the world flows by her in a stream, that she likes observing it and discovering the different characters which compose it. But in discovering these characters, she takes most violent likes and dislikes. Mrs Stock calls her a being of Imagination, she has so many ideas and such power of communicating them.'[46.]

This may allude to the streams of visitors coming to Lyme to see her and depicts her as a shrewd observer of human nature, amused by human foibles. It also suggests a certain detachment, even a sense of isolation, perhaps as a result of being unmarried, being always at Lyme – provincial, apart from the geological debates taking place in London, Oxford and Bristol – and not quite fitting into any social milieu any more. However, in taking 'violent likes and dislikes,' perhaps she was not quite so detached. The comment by Mrs Stock, who had known Mary Anning since she was a young girl, suggests

Anning was charismatic and articulate in conversation, an indication of her intelligence.

> 'Associating and being courted by those above her, she frankly owns that the society of her own rank is become distasteful to her, but yet she is very kind and good to all her own relations, and what money she gets by collecting fossils, goes to them or anyone else who wants it … She was very good-humoured with me, but gossiped and abused almost everyone in Lyme, laughing extremely at the young dandies, saying they were things or numskulls, not men. I feel exactly as she does with regard to the world at large, but have neither the impudence or the honesty always to own it. I learnt a great deal from her with regard to fossils and geology, a study I am sure I shall especially like.'[47.]

Anning seems to have been most comfortable with those of higher social strata. Her friends included men and women of very different backgrounds to her own – such as Henry De la Beche, William Buckland, Elizabeth Philpot, and Charlotte Murchison. She confided readily in some of her younger female friends like Frances Bell and Anna Maria Pinney, with whom she was also happy to gossip. Pinney records that Anning was kind and generous to her family and the needy. She may no longer have had much in common with her working class neighbours, but she never turned her back on her humble beginnings and when she did have the resources to help others, she did so. Indeed her philanthropy might go some way to explaining her fluctuating finances. Was she being over-generous to the extent that she occasionally left herself short? One anecdote connected to smuggled contraband is telling:

> 'She were good to the poor, she were; whenever she found a little cag [keg] on the shore, she would cover it up, and not let the preventative men see it, but would tell some poor person of it.'[48.]

AN ANNING ROMANCE?

An entry in Pinney's journal for 1832 hints intriguingly at an illness which, by this time, had afflicted Anning for eight years and was connected with some serious personal disappointment, betrayal or deception, perhaps a romantic attachment which came to nothing but which she thought would lift her from her poverty:

> 'I felt the power of the motions by which she was actuated, and I should have been glad to have possessed sufficient strength of mind to have done the same. An illness of eight years could not bend that spirit, although acute pain supplied the place of health, the bodily anguish was small with what must

have been suffered by a proud mind, who had hoped from childhood to see herself removed from her low situation in life, and suddenly saw those hopes blasted by Satanic treachery ... She is not quarrelling with what she cannot obtain, the season of worldly happiness once would have returned again, and I believe would even do so now, but she is wise. Her wildness of manner and late horror of everything in the world is taken for madness by those who do not understand ... But her pride is still unchastened, and she feels a kind of devilish pleasure in telling people how she hates and despises them. She even likes exulting over those whom her sense teaches her to respect ... But ... she is gentle, attentive and has a simple ... way of expressing her affectionate feelings, beyond any person I ever met with ... she will attend the sick poor night and day, when they are ill with infectious diseases, she has supported her mother and brother in bitter poverty and [even] when she was so ill that she was ... brought fainting from the Beach ... Mary says [Mrs Oakes] went and gossipped about me to old Mrs Anning to know whether I had ever told Mary that I was engaged to marry any man. The old woman replied, her Mary had no sweethearts, and she did not suppose that she went talking about that there nonsense to Miss Pinney ...'[49].

These are perceptive thoughts for someone who had just turned twenty. Anna Maria Pinney and Mary Anning were clearly sharing confidences which even Anning's mother was unaware of or unwilling to admit. Pinney paints a portrait of a woman capable of truculence and hostility, but at heart affectionate and caring. So had the kindness fallen victim to anger as a way of creating a barrier to her true feelings, the scars left by a relationship that had turned sour?

Apart from George Cumberland's 1820 gossip of a liaison between Mary Anning and Colonel Thomas Birch sparked by his philanthropy, there is no other hint of a romantic attachment: not in Mary's own letters, or those of her contemporaries. If the eight years mentioned by Pinney are correct, then we are dealing with the mid 1820s, when Anning was also in her mid twenties. The implication contained in the phrase 'who had hoped from childhood to see herself removed from her low situation in life' suggests that Mary had fallen for someone far removed from her own working class background. But she had certainly not forgotten, or forgiven – if 'hopes blasted by Satanic treachery' is anything to go by.

Potential suitors who immediately come to mind are William Buckland and Henry De la Beche. Buckland married Mary Morland in December 1825. An illustrator and fossil collector, she bore him nine children in a long and happy marriage during which they often worked together. A more likely candidate is Henry De la Beche, who returned from Jamaica in December

1824 to a personal life that was already in turmoil.

When younger, De la Beche's roving eye earned him a place in a satirical poem called *The Lymiad. A poem in the form of letters from Lyme to a friend at Bath written during the autumn of* 1818. *The Lymiad* mocks social life in Lyme and describes many of its personalities. Although unattributed, Hugh Torrens has shown that the author was the widowed Charlotte Jane Skinner, writing to a cousin in Bath. The poem calls De la Beche *Sir Fopling Fossil* and suggests that he is more interested in the young fashionable women he likes to take for outings on his yacht than in its safe navigation.

> 'He guides the helm, whilst by his side
> A damsel young and passing fair
> Reclines: her beaming eyes declare
> How dear to her Sir Fopling's smile'[50.]

However, in November 1818, in Clifton, De la Beche married the seventeen-year-old Letitia Whyte, the Irish-born daughter of the proprietor of a hotel he stayed in when in Bristol. A year later in Geneva, during an extended tour of the continent, Letitia gave birth to a daughter, Elizabeth, known as Bessie.

In 1822, De la Beche took out a lease on a large, ten-bedroom villa in Lyme called The Grove, on what is now Pound Street (then St Michael's Street). The following year he left his wife and daughter in Lyme and set off for Jamaica to visit to his estate.

Once back in Lyme, De la Beche discovered that his wife had begun an affair with Major-General Henry Wyndham, an illegitimate son of the 3rd Earl of Egremont. In 1825, De la Beche and Letitia separated and she set up home in Mayfair in London at Wyndham's home, becoming his wife in all but name. Bessie remained in Lyme with De la Beche, being cared for by his mother and stepfather whenever he was away. In 1826, De la Beche and his wife divorced in the Ecclesiastical Courts. This was not a full divorce in the modern sense; it left neither party free to remarry; that would have required a Private Act of Parliament. De la Beche never pursued this option, possibly because of its cost and the attendant publicity.[51.]

So in the mid 1820s, for the best part of two years, De la Beche was living in Lyme in the aftermath of a failed six year marriage. A footnote to *The Lymiad* says of him that talk of anything except fossils and 'the youthful geologist will take a comfortable nap in his chaise-longue.' One person in Lyme content to discuss fossils was Mary Anning.

The failure of his marriage took its toll on De la Beche's health, and he spent much of the next few years travelling in Europe, only occasionally returning to Lyme – or was that only part of the reason and was he also trying to distance himself from Mary Anning? Is it coincidence that both De la Beche and Anning were suffering from an 'illness' related to the breakup of a relationship? That their friendship blossomed into a love affair, at least on Mary's part, is entirely conjectural and based on the flimsiest of circumstantial evidence. Subsequent mentions of De la Beche in Anning's letters to her friends suggest no feelings of resentment or animosity.

Nevertheless, these tantalising hints that Anning formed a short-lived relationship have inspired two novels in recent years, Tracy Chevalier's *Remarkable Creatures* which links Mary Anning romantically with Colonel Birch and *Curiosity, A love story* by Joan Thomas who has Mary involved with Henry De la Beche. A 2020 film, *Ammonite*, directed by Francis Lee, takes a different approach and places Anning in a same-sex relationship with Charlotte Murchison. As with her putative romance with a man, there is no evidence of any with a woman. Although Mary Anning's achievements were remarkable in their own right, at times it seems she is doomed to be endlessly blown hither and thither by the ever-changing tides of faith, identity, science and sexuality. [52]

In a society where women were dependent on men, there was a social stigma attached to remaining a spinster. In her novels, Anning's near-contemporary Jane Austen stressed not only the importance of marriage, but of the need to marry well – though she herself never married. Some thought 'better any marriage than none at all'. Spinsterhood was socially acceptable only if the woman had sufficient wealth to support herself and her household. Anning's awareness of the inequalities of gender, class and wealth within the society in which she lived is shown by a quote from the Lichfield poet Anna Seward which Anning carefully copied into one of her notebooks:

> 'Nothing but an independent fortune can enable an amiable female to look down, without misery, upon the censures of the many, and even in that situation their arrows have power to wound, if not to destroy peace ...'[53]

She follows the Seward quote with a verse from a volume whose title may have been the attraction, *The World before the Flood* by the Scottish poet and hymnist James Montgomery. Interestingly, she chose to copy not an extract from the title work, but one of its other 'occasional pieces', the first verse of a sonnet entitled 'To a Bride':

> 'The more divinely beautiful thou art
> Lady! of Love's inconstancy beware;
> Watch o'er thy charms, and with an angel's care,
> O guard thy maiden purity of heart:
> At every whisper of temptation start;

The lightest breathings of unhallow'd air
Love's tender trembling lustre will impair,
Till all the light of innocence depart.'[54.]

ANNING AS A WOMAN GEOLOGIST

So much has been written about Mary Anning it is easy to form the impression that she was the only female geologist of the early nineteenth century. In fact a number of women were actively contributing to the new science at that time. Of the 237 fossil collectors who provided specimens to the Sowerbys for *The Mineral Conchology*, twenty-eight were women. Most, like Charlotte Murchison and Mary Buckland, were experienced, skilled and knowledgeable fossil collectors and artists in their own right, but worked in a supportive role, as assistant to their geologist husbands. Historian Cynthia Burek describes them as 'hidden behind husbands, brothers, fathers and colleagues, forbidden by society to expose their ability and knowledge'. Their association with their geologically-active husbands gave them an intellectual freedom and contact with other geologists denied to other women.[55.]

There were a few who made their own independent contributions and collections: Anning's friend Elizabeth Philpot and her sisters made their fossil collection accessible to researchers such as Buckland, Owen, and Agassiz; and Ann Congreve provided specimens to De la Beche and Conybeare. Female fossilists further afield included Mrs Sarah Tylee of Devizes in Wiltshire, and Mrs Elizabeth Cobbold in Suffolk, both of whom sent fossils to the Sowerbys and had fossils named after them, and the collection of Barbara, Marchioness of Hastings was purchased by the British Museum. In Scotland, Lady Eliza Maria Gordon Cumming was collecting fossil fish and corresponding with Buckland, Murchison and Agassiz. In Ireland, Mary, 3rd Countess of Rosse had an interest in geology as well as other sciences and was collecting and purchasing minerals and fossils.[56.]

But perhaps the most notable female geologist of the time, after 'the Lioness of Lyme', was another spinster, Miss Etheldred Benett who lived in Norton Bavant near Warminster in Wiltshire. Twenty years older than Anning, the independently wealthy great-granddaughter of an Archbishop of Canterbury, Benett had built up an extensive collection of mainly Wiltshire fossils and was already sending specimens to the Sowerbys when Anning was finding her first ichthyosaur. She was related to a well-known family in Lyme and certainly would have been aware of Anning's discoveries. The bulk of her collection, which she catalogued by strata, is in the Academy of Natural Sciences in Philadelphia and includes some of the first fossils found with soft tissues preserved.[57.]

There is one clear difference between all of these female geologists and Mary Anning: social class. All were middle or upper class, gentry or aristocracy while Anning was most definitely working class. Despite their wealth, their freedom was often restricted by social convention. It was socially unacceptable for them to don a pair of stout boots and ramble alone along the shore in search of fossils without either their husband or a chaperone. Anning was blessed with the freedom to go out collecting on her own or with companions of either sex. In general, these women were limited to collecting on their own estates or in their immediate local area where their social standing was known. They were women of leisure, whose collections had begun as a hobby. Even as a teenager Anning was collecting to put food on the table. Unlike Etheldred Benett, who has been described as the 'first lady geologist', Mary Anning can be regarded as the first female professional palaeontologist.[58.]

Women could not at that time indulge their geological interests to the same extent as men. Women were not admitted to membership of the Geological Society, the main forum for geological discussion in the first half of the nineteenth century, until 1919. Only the Geologists' Association admitted women members from its foundation in 1858, unfortunately over a decade too late for Mary Anning. The Geological Society did, however, publish a paper by a woman as early as 1824, when the travel writer Mrs Maria Graham wrote to Henry Warburton, then Vice President of the Society, describing the effects of earthquakes in 1822-23 she had experienced while in Chile. This was an exception; it was almost forty years before another was published.[59.]

Why are there no papers from Mary Anning in the Geological Society's published *Transactions*? Although secure and confident in her knowledge of the fossils she was discovering and preparing and with a good grasp of their anatomy, she might not have been sufficiently confident in her ability to express her thoughts for publication. The logistics of submitting a paper to a society of which she was unable to become a member was one barrier. Another was her inability to examine comparative material being found elsewhere, such as specimens from Somerset or Yorkshire. A third issue was time, of which there was never enough. In addition to her physically demanding outdoors work searching and finding fossils, cleaning and preparing them at home, and running her shop, she had her extensive correspondence to deal with, which could at times be almost overwhelming. She begins a letter of 13 January 1840 '... I am almost ready to cry at the heap of letters lying before me ...'.[60.]

This is not to say that the idea of writing a paper did not occur to her; while laboriously transcribing Conybeare's 1822 paper *Additional notices on the fossil genera Ichthyosaurus and Plesiosaurus*, Anning was clearly frustrated by his longwindedness and, no doubt delighted to have finished her task, scribbled along the bottom of the final page 'when I write a paper there shall not be but one preface.'[61.]

As early as 1829, the Bristol fossil collector George Cumberland highlighted the female contribution to geology and their lack of recognition – if in a paragraph as longwinded as any of Conybeare's

> '... but we ought ever to remember, that the world would to this day have remained ignorant of the treasures England possessed, but for the patient labours of three female pioneers in this service, viz. Mary Anning, a dealer, Miss Congrieve, and Miss Philpots, residents, who for years had been collecting and preserving these bodies from the wreck of the coast; the last two without any other view than the gratification of a laudable curiosity, and who, with unequalled liberality, communicated their collections to every man of science that visited the place; and it is to liberal minds like theirs and Miss Bennet's of Wiltshire, that we owe the first rescuing these natural gems from the spoilers ... They, and a few others, gathered the materials of this fabric raised to fame, and are entitled to a full share of the honours reaped by those who, without their aid, could never have brought them before the world.'[62.]

Each generation tends to find what it wants in the story of Mary Anning: the Victorian tract writer praises her hard work and devout character, while twentieth century and later authors find a woman fighting for a place in a patriarchal society. Increasingly, modern authors on Anning have taken to viewing her as something of a feminist icon, and rightly so. That this view is something with which Anning might have sympathised is supported by an early 'feminist' piece from an as yet unidentified source amongst the religious and moral extracts in her only surviving notebook. Entitled *Woman!* it argues against the use of scripture to repress women:

> 'And what is a woman? Was she not made of the same flesh and blood as lordly Man? Yes, and was destined doubtless, to become his friend, his helpmate on his pilgrimage but surely not his slave, for is not reason hers? Are not her claims 'To share redeeming love' as great? . . . Woman seems throughout the sacred scripture . . . more than even man the object of this pure benevolence. And woman (when his own disciples fled and left him) dared attend his cross; they were his constant followers; And women, too, were honored with the message given by the bright ambassadors of Heaven, for in the hallowed tomb, the angel spake and bade them hear the wondrous news to Peter and the rest of the Apostles, the tidings of the Resurrection! Say then shall woman sink beneath the scorn of haughty man? No let her claim, the hand of fellowship; and let her strive to prove her claim ...'[63.]

EIGHT

Fossil Fish and Financial Calamity

THE AUTUMN AND WINTER OF 1833 saw two events which further demonstrate the perils faced by Mary Anning when going about her daily business. A letter to Charlotte Murchison on 11 October, she opens with sad news:

> 'I would have answered your kind letter by the return of post, if I had been able. Perhaps you will laugh when I say that the death of my old faithful dog quite upset me, the Cliff fell upon him and killed him in a moment before my eyes, and close to my feet, it was but a moment between me and the same fate'.

Rockfalls were a professional hazard, but they were also the source of freshly exposed fossils. Anning knew the cliffs around Lyme as well as anyone, and would certainly have been aware of where a fall was likely. Yet the letter is a reminder of the risks she ran, and of her affection for her dog. Some years earlier, when laboriously copying out by hand a paper about belemnites published in the *Transactions of the Geological Society*, she was distracted momentarily by the sight of her dog curled up on the floor. Taking a few moments from her tedious task, she drew a quick sketch, first in pencil and then in ink, in her sheaf of writing paper. It is its successor, Tray, which can be seen lying at her feet in the famous portrait of Anning painted in 1842.[1].

Even in town there were dangers. At the bottom of Broad Street the road crosses the little River Lim at the Buddle Bridge. Today, the road is wide enough for two vehicles to pass but only thanks to the demolition of buildings on the south side in 1913. In Anning's day, the Buddle Bridge was just 7 feet 6 inches (2.3 m) wide. In a letter to Mary Buckland on Monday 9 December, Elizabeth Philpot gave her news that Mary Anning

> 'yesterday ... had one of her miraculous escapes in going to the beach before

sun rise and was nearly killed in passing over the bridge by the wheel of a cart which threw her down and crushed her against the wall. Fortunately the cart was stopped in time to allow for her being extricated from her most perilous situation and happily she is not prevented from pursuing her daily employment.'[2.]

Anna Maria Pinney records the same incident in her 1833 journal, although it may refer to the cliff fall which killed Anning's dog:

'8 Dec. Sunday ... Had a delightful talk with M Ag [Mary Anning] yesterday, the Word of God is becoming precious to her, after her late accident, being nearly crushed to death. I found it healing her mind as the balm of Gilead.'[3.]

DISSENTER TO ANGLICAN

Several times in her journals, Pinney refers to Anning's faith and piety. In one entry she describes how the body of a woman was washed ashore near Lyme following the wreck of the East Indiaman, the *Alexander*, off Portland in March 1815 and how sixteen-year-old 'Mary untangled the seaweed which had attached itself to her long hair, and performed all the other offices due from the living to the dead, and the unknown corpse being deposited in the Church until some friend appeared to claim it, she daily went to strew fresh flowers over it.' Although Pinney does not say so explicitly, the implication is that it was Anning who found the body.[4.]

Brought up as a Dissenter, Anning drifted towards the Anglican church as an adult. The pastor she knew as a child, James Wheaton, died suddenly in 1818 and his replacement, the Reverend John Gleed from the Independent Church at Teignmouth, 'an amiable, good-natured man of moderate ability' was much less popular. During his ten year ministry, he presided over a diminishing congregation and defections to the Anglican Church and the Baptists. Gleed had a large family to support and ran a side business importing coal. He was an enthusiastic fossil collector and may also have dealt in them. In 1827 he found a well-preserved ichthyosaur skull which was later acquired by Thomas Hawkins either through direct sale or through an intermediary, possibly Anning. Gleed's congregation felt that his business interests 'interfered with his ministerial usefulness' and in July 1828 he left Lyme for Seaton in Devon, before moving to Canada in 1832.[5.]

In 1833, a new vicar was appointed to Lyme's parish church, St Michael the Archangel, the Reverend Frederic Parry Hodges who remained its incumbent until his death in 1880. He had little time for Dissenters but openly welcomed defectors. Mary Anning and her brother probably joined his congregation in the late 1820s or early 1830s when dissatisfaction with

Gleed and his successor was at its height. It was a natural, and perhaps inevitable, transition fostered by her friendship with fellow geologists who were also Anglican clergy – Buckland, Conybeare and Sedgwick – and by her friends such as the Philpot sisters in Lyme and Sarah Kennaway and the Reverend Thomas Hodges in Charmouth. We know of the hymns and psalms which Mary Anning and the Lyme congregation sung at Sunday worship throughout the year from a hymnal complied by Parry Hodges in 1839. It includes a number of Christmas, Easter and other hymns and psalms still popular today. Anning's Christian faith strengthened as she grew older and she seems never to have had any issues reconciling the significance of her fossils discoveries with her beliefs.[6]

The same year that Parry Hodges came to Lyme, Anning met another clergyman, the Reverend Henry Williams Rawlins, Rector of Fiddington in Somerset. The Rawlins family was on holiday. Walking down Broad Street, their six year old son, Frank, was attracted to the display of fossils in the window of Anning's fossil shop. Meeting Anning, who seems to have had an affinity with children, sparked in the young boy an interest in fossils. She wrote labels for his specimens, either for some purchased or for fossils he had collected himself, and she seems to have set him thinking critically about their origin. A family memoir published in 1962 recounts the tale of young Frank, his curiosity aroused by his visit to Anning's shop, asking his father how the Genesis story of Creation explained the occurrence of fossils in rock layers at differing depths below the surface. The evangelical Reverend Rawlins' reply that the animals had all perished in Noah's Flood did not satisfy the boy who suggested instead, to his father's displeasure, that perhaps the rocks had accumulated over a long period of time after the animals had died. Frank grew up to be the Reverend Francis John Rawlins, educated at Cambridge where he would have encountered the Reverend Adam Sedgwick, and succeeded his father at Fiddington, as a more open-minded rector. His fossils from Lyme with their labels by Anning were given to the museum in Weston-super-Mare in 1913 after the death of Frank's widow but were lost by the 1960s.[7]

LOUIS AGASSIZ AND THE PHILPOTS' FOSSIL FISH
Views like those of Frank's father were widely accepted, indeed were by far the most common amongst the general population. Similar beliefs were held by a rather more influential geologist, Louis Agassiz. Son of a Swiss Calvinist minister, Agassiz viewed the fossil record as an act of Divine creation, and later in the nineteenth century, as professor of zoology and geology at Harvard University, strongly opposed Charles Darwin's ideas on evolution.[8]

In 1833, Agassiz published the first volume of his *Recherches sur les poissons fossiles* (*Research on Fossil Fish*), and came to Britain the following year to examine collections of fossil fish. To his surprise, he discovered that there were nearly 250 new, undescribed species of fish from the British strata. He visited the Bristol Institution to examine their collection where he found thirty new species. These were sent to London for him to study, along with Mary Anning's *Squaloraja* which was joined, temporarily at least, by its tail from the collection of the Misses Philpot in Lyme. Agassiz visited Lyme in October 1834 and was impressed by the Philpot collection:

'Knowing that so many fossils of all classes of the animal kingdom, and such curious plants, had been found at Lyme Regis, I was prepared to discover in this locality a greater number of fossil fishes that had been reported before a thorough search had been made. The result of my search surpassed all expectations. At Lyme Regis itself, I saw in the collection of Miss E. Philpot thirty-four new species of fossil fishes from this locality alone, several of which belong to new genera, not counting all those that I had already seen in other collections and which I found again here. Being able to stay at Lyme Regis only a few days, Miss Philpot kindly consented to allow me to borrow all these species so that I could describe them in detail. This collection has been especially valuable to me as Miss Philpot and Mary Anning have been able to show me with certainty which are the ichthyodorulites that correspond to the different types of teeth.'[9.]

The fossil collection of the Philpot sisters was by now well-known. *A list of geological and mineralogical collections in Great Britain and Ireland* published in 1829, included, under 'Dorsetshire – Miss Phillpotts, *Lyme Regis* (lias fossils); Miss Anning, *Ditto*, (specimens for sale)'.[10.]

The 'ichthyodorulites' which Agassiz mentions are the bony spines which support the dorsal fins of sharks. Anning and Philpot showed Agassiz which fossil spines corresponded to which fossil teeth, a demonstration of their expertise in field collecting, noting and recording this association. The Philpots sent many of their specimens to London where Agassiz had been given space at the Geological Society's apartments at Somerset House to work on fossil fish he borrowed from collections around the country. Some of the Philpot specimens were amongst the best examples he had ever seen. One he named *Eugnathus philpotiae*, in honour of Elizabeth Philpot:

'In dedicating this species to Miss Philpot, I have pleasure in publicly recognising the services which she has rendered to palaeontology and notably to fossil ichthyology by the care she has taken in collecting the fossil remains of the Lias at Lyme Regis. The species which we have just described and which is in her collection can be regarded as one of the finest fishes of this formation.'[11.]

Mary Anning was similarly honoured by having two species of fossil fish after her. One, a group of fish teeth from Lord Cole's collection, he named *Acrodus anningiae,* in 1843, while another he named *Belenostomus anningiae.*

A THIRD FINANCIAL CRISIS

Anning's finances had always been precarious. Mary and her mother survived seriously lean times only through the assistance of their friends like Birch, Buckland and De la Beche. The early 1830s started well, with help from sales of *Duria antiqiuor,* of her new *Plesiosaurus* to Lord Cole, and of specimens to Thomas Hawkins; within a few years Anning had built up savings of about £200. In 1833, sadly, her savings were lost, as the *Dorset County Chronicle* and other papers reported in 1836:

> 'We copy the following from a London Paper:- Mary Anning a female in humble life, residing at Lyme Regis, having by great industry and perseverance attained considerable proficiency in the science of Geology, was lately, by the sudden death of a gentleman to whom she had entrusted without receiving any acknowledgement, a small property of about £200, the fruits of her savings, to invest for her in the most advantageous manner, reduced to straitened circumstances, while her health was impaired from the hardships which she had exposed herself, and the distress of mind consequent upon her loss.'[12]

This was a serious blow. Anning was now in her mid thirties and living with her seventy-year-old mother. For the first time in her life, she had set aside sufficient savings to help secure her future. Whether she was the victim of a confidence trickster, or – as the newspaper suggests – the man to whom she had handed over the £200 died unexpectedly before her savings were invested, is unclear, but the absence of a receipt suggests that she was deliberately defrauded.

Once again, her friends and the geological community led by William Buckland rallied round to assist. Elizabeth Philpot and Sarah Kennaway appear to have initiated some sort of fund-raiser. In June 1835, Philpot wrote to Buckland: 'Mrs Kennaway has requested me to distribute a few of these papers amongst the friends of Mary Anning; it is done without her knowledge. I hope that you and Mrs Buckland will approve of it and assist the cause.' Unfortunately we have no more detail than this, so what these papers were is unknown.[13]

With a meeting of the British Association for the Advancement of Science coming up in Dublin, Buckland, unable to attend himself, wrote to the Association's General Secretary, William Vernon Harcourt in August 1835 to 'pray remember the begging box for Mary Anning's annuity'. The contributions came to £200, the same amount as she had lost. George Roberts noted later in an annotated copy of his 1834 *History and antiquities of the Borough of Lyme Regis and Charmouth* that Joseph Anning had a list of the subscribers. Buckland, with support from William Pinney, Lyme's MP and Anna Maria's elder brother, and probably from Sir Philip Egerton MP, a client of Anning's who had purchased an ichthyosaur from her in the spring of 1835, and other Fellows of the Geological Society, then persuaded the new Prime Minister, William Lamb, Lord Melbourne, to add an additional £300. As *The Manchester Times* reported:

> 'A number of the most distinguished members of the Geological Society, Messrs. Lyall and Murchieston, Drs. Buckland and Bostock, Colonel Sykes, &c., interested themselves very much for this meritorious individual, and represented her case to Lord Melbourne, with a view to obtain a small pension for her. His lordship found on inspecting the pension list that it was full; but considering that Mary Anning is one of those individuals on whom it was intended that the public bounty should be bestowed, he ordered £300 to be granted to the trustees to be disposed of in the manner which they might conceive best adapted for securing to her some provision for her declining years.'

It wasn't a fortune, but it provided Mary Anning with a modest annual income of £25, approximately twice that of a housemaid.[14.]

SALES TO SEDGWICK

In the summer of 1835, Mary Anning was once again in correspondence with Adam Sedgwick in Cambridge about the purchase of specimens for his museum. The geological collections at Cambridge University are today housed in a purpose-built museum which opened in 1904, but its foundation lies with the bequest of the large collection of a London physician, Dr John Woodward in the early eighteenth century. The bequest also provided funds to establish a Woodwardian Chair of Geology: Adam Sedgwick was the seventh Woodwardian Professor, taking up his post in 1818. During Sedgwick's tenure, the geological collections grew significantly, and he ultimately persuaded the university Senate to build a new library and museum, into which the geological collections moved in the early 1840s.[15.]

In anticipation of the new building, Sedgwick set about filling it. Using her Charmouth friend, the Reverend Thomas Hodges, once a fellow student with Sedgwick at Trinity College as her messenger, Anning listed the specimens she had for sale in a letter of June 1835:

'a perfect Ichthyosaurus about four feet and a half long, the head Vertebral column (excepting 3 of the caudal Vertebrae) in the most perfect order, the sternum as perfect as if just skinned ribs and intercostal ribs in the most perfect order bones of the pelvis good but the posterior paddle is not quite perfect else it is a picture.'[16.]

She went on to describe two other ichthyosaurs, one 'nearly as perfect except the intercostal ribs and only a side view of the sternum. About three feet in length from the tip of the nose to the tip of the tail,' and another, a large specimen perhaps 30 feet (10 m) long, still in the process of extraction: 'I have already got 80 of the Vertebrae which makes about 14 feet the Occipital Magnificent Corracoid 13 inches across Scapula a foot and a half in length ribs exceeding a yard in length.'

She also had news that she had

'within the last week or two discovered a new pentacrinite in the Oolitic sandstone bearing a general resemblance to the pentacrinites Briarus, excepting that after the arms set off from the pelvis being five they divide exactly into ten arms and no more – the largest head not exceeding a crown piece when spread open.'

This letter, with its use of osteological terms – the occipital, coracoid, intercostal ribs – demonstrates Anning's detailed knowledge of the skeletal anatomy of ichthyosaurs. Her description of the crinoid as a new species means she at once recognised its differences from a known species, *Pentacrinites briareus*. Her expertise is also apparent in a paper published by Sir Philip Egerton in the *Transactions of the Geological Society* in 1837. The previous year, Egerton had purchased an ichthyosaur specimen from her in which he was surprised to find two of the neck bones at the base of the skull fused together. Anning explained that this was always the case, and that the bones were never found separated. From this Egerton concluded that ichthyosaurs had powerful necks, but limited movement of the head. It is this level of detailed knowledge and expertise that marks Mary Anning out. She was now well beyond selling mere 'curios' as souvenirs, but was providing an expert service to museums and collectors.[17.]

Nowhere in her letter to Sedgwick does she mention prices. Sedgwick must have replied on 24 July asking for them. Three days later Anning told him that she wanted £50 for the four and a half foot ichthyosaur, and £20 for the smaller, describing them as 'the most perfect yet discovered'. She seems to have had a spell of good fortune, with still more ichthyosaurs coming to light. Her letter explains that she is still excavating others, one about 12 feet (4 m) long, evidently from low on the shoreline as the tide 'would not allow of our working above one hour in a day,' and a second one which she describes as 'the smallest I have yet seen about 1 foot 9 inches in length'; unfortunately 'the body part is enveloped with pyrites but as to general form good as to the head and tail they are exquisite'. Pyrite, iron sulphide, 'fool's gold' which is common in the Lias rocks, is susceptible to decay in damp conditions so fossils preserved in pyrite or enclosed in pyrite can slowly self-destruct, even in a museum storeroom.[18.]

Sedgwick bought the £50, four and half foot ichthyosaur, and soon it was on its way to London by sea, as Anning informed him:

'I sent off the Ichts on tuesday 2nd of September on board the Unity, Pearce Mastr which I hope er'r this arrived safe and I trust you will not be dissapointed when you Sir see it, whilst packing it I had the pleasure to discover the greater portion of the second posterior paddle, which previously was the defect I mentioned in the skeleton.'[19.]

Once again, the *Unity* under its master Robert Pearce was Anning's vessel of choice when shipping specimens from Lyme. On 23 September, Anning acknowledges receipt of the £50. The specimen, which survives in the Sedgwick Museum, seems to have been a personal purchase by Sedgwick who then presented it to the university collection.[20.]

The large specimen mentioned in her letter of June 1835 to Sedgwick may be that which was reported in the press in April:

'Ichthyosaurus. – A letter from Mr. Roberts, the intelligent historian of Lyme Regis (See *Lit. Gazette*, No.944), informs us that the extraordinary female geologist of that part of the coast, Miss Mary Anning, has discovered the largest Ichthyosaurus ever found. "The gigantic animal (he says) must have died, and its bones fallen abroad at the decomposition of the body just before they were covered with lias deposit, which became a layer of lime-stone. The bones lie as usual in the marl between. This animal presumed to be the skeleton of the Ichthyosaurus Platyodon, must have been at least thirty-five feet in length and of considerable bulk. The great one, recently purchased, at the British Museum,* was only half as large. Miss Anning can get but a few bones at a time, as the ledges are broken up, so scattered are the bones."
'*This is a sore subject. It might have been six times as large!! – *Ed. L.G.*'[21.]

The editor's note here probably relates to the then current controversy over the plasterwork on the Thomas Hawkins' specimens recently purchased by the British Museum.

In August 1837, Mary Anning was visited by the Prussian naturalist Ludwig Leichhardt who was continuing his studies in England and touring the West Country. As he recorded in his letters:

'Later on, we walked to the coast of Devon and stayed at Lyme Regis for 8 days. We saw the dark walls of the Lias against which the sea beats daily, dislodging the vestiges of the primeval world. We walked over thousands of ammonites embedded in the smooth-washed, slippery paste of the Lias along the shore. And we had the pleasure of making the acquaintance of the Princess of palaeontology, Miss Anning. She is a strong, energetic spinster of about 28 years of age, tanned and masculine of expression. Every morning, and after every stormy sea, she goes walking and clambering about on the slopes of the Lias to see whether fossils have been brought to light by falls of rock or wave action. Naturally, she is collecting for sale, and her shop contains nearly all the fossils of the Lias.'[22.]

Anning's outdoor life must have kept her looking young: she was ten years older than Leichhardt estimated. Leichhardt later went on to explore the interior of Australia, and disappeared on an expedition in 1848.

In 1838, Anning was celebrated in verse by the poet and philanthropist John Kenyon who had visited Lyme two years before and 'fossiled with that very interesting person Mary Anning'. *To Mary Anning,* published in Kenyon's 1838 *Poems: for the most part occasional,* was inspired by her surviving the lightning strike and, seemingly, a near-drowning, and focuses on her childhood discoveries. In a footnote he writes:

'Miss Anning, of Lyme Regis, is advantageously known as the earliest and most skilful explorer of the Saurians, and other fossils remains, in that neighbourhood. The most eminent geologists have been forward to attest the value of her discoveries. Those who know her personally, will be no less eager to bear testimony to her kindly temper, her straightforward character, and her fresh and strong intellect. The first line alludes to her having been struck senseless by lightning when a child. She was restored with difficulty; but the nurse, in whose arms she was, and two other females, were killed outright by the same flash.'

As Anning historian Tom Goodhue has pointed out, Kenyon rather limited his options by ending each of his poem's six stanzas with 'Mary Anning', the result being, according to Goodhue, 'truly dreadful'. The first and second verses are sufficient proof of its quality:[23]

> Thee, Mary! first 'twas lightning struck,
> And then a water-vat half drowned;
> But I can't think 'twas mere blind luck
> Twice left for dead—twice brought thee round.
> No! Fortune in her prescient mood,
> I must believe, e'en then was planning

> To fabricate a something good
> Of Thee, the twice-saved Mary Anning.
>
> This to fulfil she did not bid
> Thy feet o'er foreign soils to roam,
> For well she knew what powers lay hid
> In these blue cliffs that touched thy home.
> And hither led, in vain to Thee
> Or marle, or rock, was insight banning;
> Some folk can through a millstone see;
> And so, in sooth, can Mary Anning.

From late in 1839, we have the only known scientific publication she wrote, an extract of a letter she sent in April to Edward Charlesworth, editor of *The Magazine of Natural History*. Although only part of the letter is published, it appears that she is responding to an enquiry from Charlesworth relating to the teeth of a fossil shark.

'Note on the supposed frontal spine in the genus *Hybodus*.

'*Extract of a Letter from Miss Anning*, referring to the supposed frontal spine in the genus *Hybodus*. – "In reply to your request I beg to say that the hooked tooth is by no means new; I believe that M. De la Beche described it fifteen years since in the Geological Transactions, I am not positive; but I know that I then discovered a specimen, with about a hundred palatal teeth, and four of the hooked teeth, as I have since done several times with different specimens. I had a conversation with Agassiz on this subject; his remark was that they were the teeth by which the fish seized its prey, – milling it afterwards with its palatal teeth. I am only surprised that he has not mentioned it in his work. We generally find the *Ichthyodorulites* with them, as well as cartilaginous bones." – *Mary Anning. – Lyme Regis, April 7, 1839.*'[24]

Earlier that year, Anning had found a fine specimen of *Hybodus* jaw which she sold to Edmund Thomas Higgins, a surgeon and fossil collector. Higgins passed the specimen to Charlesworth for description and publication and although Charlesworth states that the fossil had been found jointly by Anning and Higgins, there is no evidence that that is the case. In fact another account of the discovery mentions 'the jaws and other fossil remains of an extinct species of Shark, discovered by Miss Anning in the cliffs of Lyme Regis and in the cabinet of Edmund Higgins Esq.'[25]

Anning's letter demonstrates her observations on the different kinds of teeth present in *Hybodus* – four hooked teeth, actually grasping spines used in mating, and a set of crushing teeth – and, again, her knowledge of the literature, in this case Agassiz's work on fossil fish.

Charlesworth's account of the fossil had been published in the May 1839 issue of *The Magazine of Natural History* and he recognised it as a new species which he named *Hybodus Delabecheii* after Anning's friend, Henry De la Beche. The next, June 1839, issue of the magazine carried a follow-up paper about *Hybodus* from the Isle of Wight and a letter from Henry Woods about a specimen in the museum in Bath. Anning's letter, although written as early as April, does not appear until the December 1839 issue, almost as an afterthought, and with a rather dismissive and patronising editorial postscript from Charlesworth:

> '[As Miss Anning speaks of 100 *palatal* teeth, she probably refers to the genus *Acrodus*, which may very possibly be furnished with an organ similar to the one possessed by *Hybodus*, as the genera are closely allied. Mr. De la Beche makes no allusion to its existence in the Geological Transactions. – Ed.]'

This obviously rankled. Anning continued her *Hybodus* correspondence with Charlesworth in a further letter, unpublished, in July 1839 in which, with more than a little false modesty and sarcasm, she states 'as I am illiterate, [I] am not able to give a correct opinion'.[26]

Higgins donated fossil fish and marine reptile fossils, some of which may have been collected by Mary Anning, to the Yorkshire Museum in the late 1840s. The specimen of *Hybodus delabechei*, found by Anning and illustrated and described by Charlesworth, is today in the collections of the British Geological Survey.[27]

In the late 1830s, Richard Owen of the Royal College of Surgeons was preparing a report on British fossil reptiles for the British Association for the Advancement of Science. Writing to his father-in-law, William Clift, in September 1839 he outlined his plans to 'spend a day with Mr Hawkins at Street and then take a run down to make love to Mary Anning at Lyme, and then post home as fast as stage-coach can carry me.' This curious turn of phrase simply means that he was intending to ingratiate himself with Anning and flatter her with a view to learning her news of any recent discoveries and perhaps seeking her advice on the Lias marine reptiles. Nothing quite worked out as he hoped. It was on this visit that Owen was waylaid by Buckland and Conybeare and taken to Axminster, and the following day the three of them with Mary Anning had to scramble to safety from an incoming tide. He closed his visit with a day spent examining the Philpot collection at Morley Cottage.[28]

DOWLANDS LANDSLIP

The 1830s ended in spectacular fashion in Lyme. After six months of rain, on Christmas Day 1839 a huge part of the coastal cliff about 3 miles (5 km) west of the town collapsed. As the cliff slid towards the sea, a chasm 100 yards wide and 150 feet deep (92 m wide and 46 m deep) opened up behind it, extending along the coast for nearly a mile. Eight million tons of cliff had moved seawards, forcing up a shingle ridge out of the sea as the toe of the landslip rose. A large detached block of clifftop, covering 15 acres (6 ha), took with it a wheat field which was ceremoniously harvested the following year.

The landslip is well-documented because the land was farmed and occupied and there are eye-witness accounts. In addition, fortuitously, both Buckland and Conybeare were soon on the scene from nearby Axminster where Buckland was visiting from Oxford with his family. Mary Anning accompanied them, along with Buckland's wife, Mary. Conybeare quickly published a description of the landslip in the local press which was widely reprinted. Onto a copy of one of those reprints, published by Daniel Dunster of Lyme Regis, Anning wrote a letter to a friend in London on 13 January 1840,

> 'I have with Buckland [and] Conybear wandered all over this splendid subsidence and I believe this printed account was written by Conybear, Buckland said it was worth coming 500 miles to see.'

Buckland's remark was prescient: the landslip quickly became a tourist attraction, with visitors, including Queen Victoria, arriving by both land and sea. Anning's interest, though, was limited as the landslip offered her no collecting opportunities, 'being only loose sand and chalk there are no fossils in it,' she wrote, disappointedly.[29.]

Lyme schoolmaster and historian George Roberts published a pamphlet on the landslip which quickly ran through at least five editions, while Conybeare and Buckland, along with an Exeter surveyor, William Dawson, published what is perhaps the earliest scientific description and analysis of a landslip, with plans and sections by Dawson and some magnificent illustrations drawn by Mary Buckland.[30.]

NINE
'I am well known throughout the whole of Europe'

By the mid 1840s, the marine reptile fossils were no longer fetching the prices they had done even in the 1830s, much less the 1820s. Gideon Mantell attended an auction in July 1844 where some of Hawkins' specimens were for sale and noted that 'none of the valuable specimens were bid for'. He bought a 'skull of a young Ichthyosaurus for 24/-.'[1]

In October 1840, the British Museum purchased three fossil starfish from Mary Anning. Recorded in the museum's catalogue as 'Species of Ophiura (O. Egertoni?) on blue lias Lyme Regis Bought of Miss M. Anning,' this is the first time that her name is formally recorded as the collector of a specimen acquired by the museum. Another purchase from Anning took place in 1846 when the British Museum bought a specimen showing two kinds of teeth belonging to the fossil shark, *Hybodus*. Other fossils found by Anning were already in the museum's collection by 1840, either purchased directly or having come to the museum via other collections such as the ichthyosaur bought at Bullock's sale in 1819 or Thomas Hawkins' specimens which the museum acquired in 1834 and 1840. In cases such as these, Anning's name was no longer associated with particular specimens and it is generally only through much research by modern curators that the link to Anning has been re-established, and sometimes only tentatively. This applies not only to the Natural History Museum which houses the majority of her specimens but also to other museums around the UK and elsewhere in the world where Anning specimens have ended up. As a commercial dealer, Anning's name was simply not recorded.[2]

HUGH STRICKLAND AND THE AMMONITE APTYCHI
Amongst the geologists to visit Lyme in the early 1840s and to benefit from Anning's sharp eye was Hugh Edwin Strickland. A student of Buckland at Oxford, and a friend of Murchison, Strickland had wide geological and natural history interests. In a paper for the Geological Society published in

1845, Strickland described a feature of ammonites that had first been pointed out to him by Mary Anning when he visited her in Lyme four years earlier. In his paper he writes,

> 'In 1841, Miss Anning, of Lyme Regis, drew my attention to some black-coloured substances which she had occasionally met with in the interior of the *Ammonites Bucklandi*, and which she considered to indicate the presence of the ink-bag in the animal of the Ammonite ... From these and other specimens, it appeared to me evident that these substances had constituted, not an ink-bag, but a laminar appendage of the animal, adapted to discharge some unascertained function ... Miss Anning informed me that these bodies generally occur about the middle of the outer chamber of the ammonite, whence they are obtained by breaking the fossil; but as this process more frequently destroyed than exposed the object of search, I was unable, during my stay at Lyme Regis, to procure any tolerably perfect specimen.'[3]

Anning had noticed within the shells of ammonites what she thought were ink-sacs similar to those she had found associated with belemnites in the 1820s. Strickland had no success in his search when he was at Lyme, but a few years later, while studying ammonites from the Lias rocks in Warwickshire, he found what he was looking for. This was a pair of small, almost semi-circular shells, similar to small clam shells, lying within the large outermost chamber of an ammonite. Strickland recognised these as plates which could have been used by the ammonite to close off the opening of the shell. Strickland's discovery shed some light on a puzzling fossil called *Trigonellites*, also called *Aptychus*, which had been known since 1811. Today we know these features of ammonite shells as aptychi, and although they may be used as shutters to close off the ammonite shell opening, palaeontologists are still debating whether they had other functions.[4]

MARY MOURNS

In October 1842 Molly Anning died, at the age of seventy-nine. Her death was noted in the *Taunton Courier, and Western Advertiser*:

> 'Oct 5, at Lyme Regis, of paralysis. aged 79, Mary, the widow of Richard Anning, and mother of the celebrated Miss Mary Anning, the fossilist.'[5]

The 1841 census listed the two Mary Annings, one seventy-seven years old, the other forty, both as 'fossilists'. Memories of a young customer, written in old age towards the end of the century, recalled

> 'Mrs. Anning the Fossil Woman's mother ... a very old lady in a mob cap and large white apron, who sometimes came with feeble steps into the shop ... But she is a very dim figure in the far distance of my recollections, yet a

portion of the view which in justice to her attached daughter must not be left out, for the two were devoted to one another.'[6.]

Like many in the nineteenth century, Mary Anning kept a series of notebooks or commonplace books. At a time when printed books were expensive the only way to keep an interesting passage was to copy it out. Only one of Anning's notebooks, her fourth, survives today. Into it she copied quotations, extracts from poems and prayers. Interestingly, there is no mention of her profession, or of fossils. This has meant that her sole surviving commonplace book has been dismissed in the past as typical of 'the sort of extracts and jottings which a young lady of the period might be expected to make,' of 'no permanent value,' and of little interest other than to 'point to the simplicity and piety of this remarkable and good-hearted woman.'[7.]

In more recent times, however, it has been recognised that Anning's commonplace book does indeed tell us much about her, that her choice of quotations is a clue to her personality and inner life. They includes passages from poems by Lord Byron, Henry Kirke White, Tobias Smollett, Thomas Gray and Hannah More and prayers by Thomas Wilson. Unfortunately the entries are undated so it it not clear when they were made or what period they cover. Several poems dealing with death, solitude and loneliness were probably entered after the death of her mother, suggesting that she was struggling to adapt to being on her own for the first time in her life.[8.]

With her mother gone, Anning carried on with the fossil business. Her small annuity of £25 helped, but she still needed the income from the sale of fossils. In 1843 she was again in correspondence with Adam Sedgwick in Cambridge about the purchase of further specimens for his museum. In reply to his enquiry about what she had available, Anning told him in a letter of early May that

> 'the platyodon Head is 4 feet and 3 inches long general contour good but the eyes are a little crushed otherwise an illustrative specimen and worthy of a place in a Museum Price £4 – Next I have a picture of an Ichts 4 feet 3 inches lying on its back the sternum as perfect as if just taken from a dissecting room and although the dorsal vertebrae are dislocated it is an advantage as showing the intestinal skin Sir I just send you a rough scratch of it price £20.'

She also had another ichthyosaur, 7 feet (2 m) long, for £10 and small slabs of crinoids for £1.10.0 and £1.5.0, and a range of other specimens – parts of fish and ichthyosaurs, ammonites, some cut to show the internal structure, a section through ichthyosaur vertebrae and ribs. The quality of the specimens, she tells him, can be attested to by the Dorset clergyman and geologist the Reverend Osmond Fisher, whom Sedgwick would have known in Cambridge, and by 'Mr Conyber who is quite in raptures with them'. Sedgwick purchased the *Ichthyosaurus platyodon* head and in a letter of 20 May, Anning tells him 'I have sent off the platyodon head for waggon railroad to London' and that the cost of the packing case was an additional seven shillings. Payment for £4. 7. 0. was acknowledged by Anning on 26 May 1843. This specimen is still in the Sedgwick Museum, although it is significantly smaller than the 4 feet 3 inches (1.3 m) indicated in her letter. It appears than Anning was confusing its size with that of the complete skeleton for which she wanted £20. This latter specimen, of which Anning provided a sketch – a 'scratch' as she calls it – was also purchased by Sedgwick, as her drawing matches precisely a specimen in the museum today, but there is no correspondence or records by Sedgwick of its acquisition.[9.]

By the 1840s, Mary Anning was in correspondence with Mrs Dorothea Solly, wife of the geologist Samuel Solly FRS, then living at 48 Upper Gower Street in London. Dorothea was the daughter of the Dorset naturalist and antiquary, the Reverend Thomas Rackett, and it seems that the Sollys and Racketts had visited Lyme, probably in the late 1830s, becoming friends with Mary Anning and the Philpot sisters. In a letter of June 1844 to Dorothea Solly, Anning, feeling isolated from the scientific centre of London, thanks her for sending her and the Philpots news of events there:

> 'be pleased to accept my most Grateful thanks; for yr still condescending to remember me, as also for yr most interesting account of the lectures (Lyme being such an out of way Town, that I seldom get any scientific news) Miss's philpots desires their kind Compts and were so delighted with the information contained in yr letter; that they fingered [borrowed] it to copy.'

Then she wrote:

> '... I can only remark on it generally as truly believing from what little I have seen of the fossil World and Natural History, I think the connection or analogy between the Creatures of the former, and present World, excepting as to size, much greater than is generally supposed.'

Anning here is noticing, based on her own experience, the similarities, other than in size, between fossil and living animals. Amongst her papers in the Natural History Museum are several pages of notes on the differences between living and fossil species copied from Conybeare and Phillips' 1822 book *Outlines of the geology of England and Wales*. Again we see Anning as more than a discoverer and seller of fossils; she is trying to understand their origins, both through her reading, and by talking to the procession of palaeontologists who sought her out when they came to Lyme.[10.]

In her letter to Mrs Solly, Anning bemoans the lack of recent fossil finds: '... as we have not had either storms or Landslips this past winter there has been but little found' but expresses her pleasure that the Sollys have 'made the acquaintence of my Old friend Sir H. De la Beach'. The way in which she refers to De la Beche here and in earlier letters suggests that he was simply the 'Old friend' with whom she had scrambled over the rocks when both were young and all life lay ahead, and not the cause of her upset in the mid 1820s.

LYME BURNING

On 11 May 1844, Lyme was ravaged by another great fire which, like that of 1803 started at a baker's premises. Preceded by two months of drought, and fanned by a strong northeasterly wind, flames swept the town, burning from ten o'clock in the morning to ten at night. Anning included a description in her letter to Dorothea Solly written a month after the event:

> '... if we have not had a Landslip we have had a tremendious fire skip viz 52 houses including 3 inns it began at the George Inn commonly called Monmouth house, the wind being very high the 2nd House sellers in broad street and cleared all that side as far as the bridge close by were you lodged skipped over yr old Lodgings and then burnt both sides of the street viz namely Comb Street, to cut off the fire in broad Street they pull down they shambles I do regret the Old Clock that had stood for Centries ...'

The fire had taken hold on the east side of Coombe Street, not far from the Independent Chapel, quickly spreading west towards the sea, burning down the Custom House and the Cups Hotel. To prevent the fire from spreading up the seaward side of Broad Street, a group of buildings called The Shambles which lay in the middle of the street was pulled down. Joseph Anning was living at Pyne House facing The Shambles, and their demolition was his saviour. Mary Anning's house and fossil shop farther uphill was not affected. Over fifty houses were destroyed or partly burnt, but no lives were lost. The fire featured in the *Illustrated London News*, accompanied by a dramatic woodcut 'in which the whole of Lyme Regis seems to be ascending in flames to heaven'.[11.]

CARUS AND THE KING

Two months after the fire, Mary Anning received a visit from Frederick Augustus II, King of Saxony, then on an informal tour of Britain accompanied by his personal physician, the naturalist and artist Carl Gustav Carus. They had arrived in England on 29 May on a whistle-stop tour from Dover to Penzance to Fort William and Inverness, and many points in between, before they departed from Edinburgh on 5 August. With his scientific interests, Carus took every opportunity to view museum collections and meet with scientists. In London he visited the Royal College of Surgeons where he met Richard Owen and saw specimens brought back from South America by Charles Darwin. Carus was impressed by the variety and number of fossils which had been found in England: 'There is ... no country which offers so many inducements and opportunities for the study of fossil remains as England,' he wrote. 'Professor Owen told me that ... the remains of single portions of at least 1000 Ichthyosauri ... had been already discovered.' At the British Museum he met Charles Konig, a fellow German-speaker, and saw what he called the 'marine Amphibia (enaliosauria),' the marine reptiles, 'which are particularly represented in the different species of Plesiosauri and Ichthyosauri, the most splendid specimens of which have been found on the sea-coast, near Lyme Regis. Our attention was especially directed to an immense spinal column of an Ichthyosaurus, quite detached, more than twenty feet long.'[12.]

On 1 July 1844, monarch and physician travelled from Weymouth along the coast and

> 'at length reached Lyme-Regis ... the scene where so many fossils exist, and particularly the beds which contain those singular fossil sea-lizards ... we fell in with a shop in which the most remarkable petrifactions and fossil remains—the head of an Ichthyosaurus—beautiful ammonites, &c, were exhibited in the window. We entered and found the small shop and adjoining chamber completely filled with the fossil productions of the coast'.

There had been few cliff falls the previous winter and the stock in Mary Anning's shop was much reduced. Nonetheless, Carus noticed that there was

> 'a large slab of blackish clay, in which a perfect Ichthyosaurus of at least six feet, was imbedded. This specimen would have been a great acquisition for many of the cabinets of natural history on the Continent, and I consider the price demanded, 15 [pounds] sterling, as very moderate. I was anxious, at all events, to write down the address, and the woman who kept the shop—for it was a woman who had devoted herself to this scientific pursuit—with a firm hand, wrote her name, "Mary Annins," in my pocket-book, and added, as she returned the book into my hands, "I am well known throughout the whole of Europe."'[13.]

This was no exaggerated boast. She was indeed well-known in scientific circles across Europe. Her fame had reached George Cuvier in Paris by 1824, and Neuchâtel in Switzerland through Louis Agassiz in the 1830s, Henry

De la Beche would undoubtedly have mentioned Anning and her discoveries while visiting European museums and scientists on his various continental tours between 1819 and 1829. She was talked about in Berlin in 1831 when the Prussian geologist Leopold von Buch lectured to the *Berliner Gesellschaft der Freunde der Humanität* (the Berlin Society of Friends of Humanity) about Henry De la Beche's newly-published lithograph *Duria antiquior.* Von Buch described the discoverer of the fossils shown in the picture:

> 'Miss Anning is a knowledgeable lady who of late has become quite renowned on the continent for her discoveries in England. She lives in Lyme Regis on the seashore of Dorset. At each low tide, she gathers up her skirts and with an astonishing perseverance searches the deep, black and rich Lias mud for objects which the waves have ripped out and washed up. Initially just for her own instruction, she now makes a profit from the pastime and successfully sells what the flood bestows on her.'[14.]

She was known in the United States also, not only through the specimens sold to museums and collectors there, but through an 1840 children's book, published both in England and in America, *Peter Parley's Wonders of the Earth, Sea and Sky.* This book, by American writer Samuel Griswold Goodrich, devotes a chapter to 'What creatures once lived where Dorsetshire in England now is' and recognises that what is known is due to 'a lady, Miss Anning, who spends nearly her whole time in collecting fossils out of the cliffs. No one ought to go near Lyme Regis without visiting her collection.' [15.]

HER FINAL YEARS

By 1845 Mary Anning was suffering from the breast cancer which would eventually claim her life. She was in considerable pain. Her worries and difficulties steadily worsened. Into her commonplace book she copied out prayers by Bishop Thomas Wilson, including his *Prayer under lingering illness* and a *Prayer to be used for a sick person* – with the lines: 'Deal thus with this thy servant, O Lord, and do as Thou knowest best for him; either relieve him in thy mercy, or mercifully enable him to bear this burthen with patience.' One source of local gossip whispered that she had taken 'to strong drinks and opium to ease the pain.' Laudanum, a solution of opium in alcohol, was a common medication at the time, so it is quite possible that she had to resort to this for pain relief.[16.]

We have an account of her at about this time, albeit a recollection from fifty years later. In 1895 Eleanor Emma Waring, the granddaughter of fossil collector Captain Henry Waring RN, remembered that

> '"Miss Anning the Fossil Woman" lived in Broad Street, in a house with a

small shop front, and which is now occupied by Mr Beer the Hairdresser, there lived this very timid, very unpretending, very patient, and very celebrated woman, the discoverer of the Icthyosaurus and of other fossil remains which were living animals before the Deluge. We, as children, had large dealings with Miss Anning, our pocket-money was freely spent on the little Ammonites which she washed and burnished till they shone like metal, and on stones which took our childish fancy. She would serve us with the sweetest temper, bearing with all our little fancies, and never finding us too troublesome as we turned over her trays of curiosities, and concluded by spending a few pence only, and this we might do as often as we liked without offence. She must have been poor enough, for her little shop was scantily furnished, and her own dress always of the very plainest, she was very thin and had a high forehead, and large eyes which seem to me to have a kindly consideration for her little customers.'[17]

This description of Anning, within a year or so of her death, as timid and very thin may reflect the impact of her cancer on her manner and appearance, but in spite of it she clearly remained patient and friendly towards her young clientele.

There were some happier moments. 1846 saw some further recognition for Anning's contribution to science. Hearing she was ill, Buckland again initiated a subscription through the Geological Society to supplement the annuity of £25 she had been granted eight years earlier. At a meeting of the Manchester Geological Society, it was announced that a subscription had been launched at that year's British Association meeting in Southampton

> 'for Mary Anning, of Lyme Regis, an old woman, [she was 47] who had for many years been employed as a collector of fossils ... Owing to age, and rheumatic affection caught whilst the thatched roof of her house was wet from having had water poured on it to extinguish a fire, she was unable to follow her former occupation of collecting fossils ...' .[18]

In July 1846 the newly-established Dorset County Museum in Dorchester elected Mary Anning as its first honorary member. Amongst the museum's earliest supporters were several of her friends, Osmond Fisher, curate at All Saints' Dorchester, and the Reverend Thomas Hodges of Charmouth.[19]

As well as copying comforting prayers, Anning also kept some more light-hearted bits and pieces. Amongst her few surviving papers is a sheet of decorative writing paper with an attractive engraving of Windsor Castle at its head. Produced by stationers and printers John and Frederick Harwood of Fenchurch Street in London, this was quality stationery. Sometime, probably late in 1846 or early the following year, Anning put it to use by copying out a humorous poem in praise of the recently-knighted husband of Anning's friend Charlotte, now Lady Murchison.

> Who first survey'd the Russian states?
> And made the great Azoic dates?
> And work'd the Scandinavian slates?
> Sir Roderick.
>
> Who calculated Nature's shocks?
> And proved the low Silurian rocks
> Detritus of more ancient blocks?
> Sir Roderick.

This poem, of which these are the first of nine verses, all in much the same vein, celebrates the new Sir Roderick Murchison's geological researches in Russia and the Silurian rocks of Wales, and also makes references to Sedgwick, Buckland (and coprolites), Agassiz and Owen, all of whom were known to Anning. On the same paper she continued with a copy of 'The Complaint of an abused Sunbeam against Dr. Faraday' but the appeal of this to Anning is less clear, as it is more about recent developments in physics than geology. Often misattributed as Anning's own compositions, the two poems were first published in a magazine called *Bentley's Miscellany* in 1846 and were written by an unknown, young Cambridge physician clearly up to date with current developments in geology and physics and using the pseudonym of 'The Travelling Bachelor'.[20.]

This copy of *Bentley's Miscellany* must have been lent to Anning by one of her Lyme friends, perhaps in an attempt to amuse and distract her as her illness worsened. As the cancer progressed and her health deteriorated, she must have been housebound. No more would she 'gather up her skirts'. Her fossil-collecting was over. Alas, the last few months of her life are a complete blank. We know nothing of how she was cared for or by whom, but that burden of care must have fallen on her brother and his wife who lived a few doors down Broad Street. After living with breast cancer for two years, Mary Anning died on Tuesday 9 March 1847.

A week later, she was buried in St Michael's churchyard, atop Church Cliffs where she found so many of her fossils, by the vicar, the Reverend Frederic Parry Hodges. We have no record of who attended her funeral, but her close friend Elizabeth Philpot – the last of the Philpot sisters, she had already laid her two siblings to rest in that same churchyard – must surely have been present. Perhaps Sarah Kennaway made the journey from Charmouth to attend. The gravestone inscription reads,

'Sacred to the Memory of **JOSEPH ANNING**. Who died July the 5th 1849

Aged 53 years. Also of three Children who died in the Infancy Also of MARY ANNING Sister of the above Who died March the 9th 1847 Aged 47 years.'

The stone was erected by Joseph's widow, Amelia, who gives her late husband precedence, and also commemorates the burial of three of their five children. The gravestone's weathered lettering has been recut several times, so the exact wording of the original is unknown.[21]

Mary Anning's death did not pass unnoticed. *The Gentleman's Magazine* carried the news of the death of 'the celebrated geologist, a delightful discoverer of the fossils of the blue lias.' George Roberts contributed an obituary to the London-based literary magazine, *The Athenaeum*. Much of it was taken up with her survival as a baby from the lightning strike, but it did describe her as 'of European fame as a discoverer of fossils'. The *Athenaeum* notice was picked up and republished by regional newspapers across Britain and Ireland, even reaching the far-flung corners of the Empire for the benefit of the readers of *Home News for India, China and Colonies* – few of whom would have known who she was.[22]

One man who did, and provided the obituary that would have most pleased her, was Henry De la Beche. His journey had been as eventful as her own, his rewards far greater. He was now Sir Henry, Director of the Geological Survey, President of the Palaeontographical Society and, in 1848, President of the Geological Society.

On Friday 18 February 1848, the Fellows of the Geological Society of London met for their Annual General Meeting at Somerset House. As had been the custom for the previous twenty years, in his Anniversary Address the President noted the passing, in the previous twelve months, of notable Fellows. In a remarkable break with this tradition, De la Beche turned his attention to someone who was not, and could not be by virtue of her gender, a member of the Geological Society:

> 'I cannot close this notice of our losses by death without adverting to that of one, who though not placed among even the easier classes of society, but who had to earn her daily bread by her labour, yet contributed by her talents and untiring researches in no small degree to our knowledge of the great Enalio-saurians, and other forms of organic life entombed in the vicinity of Lyme Regis. MARY ANNING was the daughter of Richard Anning, a cabinet-maker of that town, and was born in May, 1799 ... From her father, who appears to have been the first to collect and sell fossils in that neighbourhood, she learnt to search for and obtain them. Her future life was dedicated to this pursuit, by which she gained her livelihood; and there are those among us in this room who know well how to appreciate the skill she employed (from her knowledge of the various works as they

appeared on the subject,) in developing the remains of the many fine skeletons of Ichthyosauri and Plesiosauri, which without her care would never have presented to comparative anatomists in the uninjured form so desirable for their examinations. The talents and good conduct of Mary Anning made her many friends; she received a small sum of money for her services, at the intercession of a member of this Society with Lord Melbourne, when that nobleman was premier. This, with some additional aid, was expended upon an annuity, and with it, the kind assistance of friends at Lyme Regis, and some little aid derived from the sale of fossils, when her health permitted her to obtain them, she bore with fortitude the progress of a cancer on her breast, until she finally sunk beneath its ravages on the 9th of March, 1847.'[23]

Amongst those in the room were Adam Sedgwick and William Buckland – at that same meeting De la Beche presented Buckland with the Society's highest honour, the Wollaston Medal. Also present, although as far as we know he never met Anning, was the Society's former Secretary, Charles Darwin.

A more lasting memorial to Mary Anning in the form of a new stained-glass window was installed in St Michael's Church in February 1850. The design was probably derived from a slightly earlier Staffordshire church by the Gothic Revival architect Augustus W.N. Pugin, with windows from the renowned stained glass workshop of William Wailes in Newcastle-upon-Tyne. The Anning window may have been a standard 'off-the-shelf' design. It depicts the six acts of mercy from St Matthew's Gospel, with, along its base, a Gothic inscription:

> 'This window is sacred to the memory of Mary Anning of this parish who died 9 March 1847 A. D. and is erected by the vicar of Lyme and some of the members of the Geological Society of London in commemoration of her usefulness in furthering the science of geology and also of her benevolence of heart and integrity of life.'[24]

The vicar was the Reverend Parry Hodges who had welcomed Anning to the Anglican church and who laid her to rest. The stained glass design – probably the selection of Parry Hodges – was particularly appropriate to commemorate Mary Anning. She was well-known in Lyme for her acts of kindness, generosity and charity towards the poor. The window was paid for, not through an official contribution by the Geological Society (for which there is no mention in the minutes of the Society's Council), but by contributions from friends of Anning who were Fellows of the Society. We have no record of who contributed but we might expect her close friends De la Beche, Buckland and Conybeare to have funded the installation of the window, and perhaps also Sedgwick, Murchison, Owen, Hawkins, Egerton and Cole. Some notes by George Roberts suggest that in addition to the window, 'some members of the Geological Society subscribed altogether £10 towards a tablet to the memory of Miss Anning'. No such tablet exists today and there seems to be no other record of it, so perhaps it was never installed or has since been lost.[25.]

Within ten years the geologists most closely associated with Mary Anning were dead. In three successive years Henry De la Beche, William Buckland and William Daniel Conybeare died: De la Beche in 1855, Buckland in 1856 and Conybeare in 1857. Anning's friend and fellow collector, Elizabeth Philpot, also died in 1857. The *Taunton Courier and Western Advertiser* reporting the deaths of 'Dean Conybeare and Miss Elizabeth Philpot, of Lyme' noted that 'Miss Philpot went upon the lias shore in company with Mary Anning almost daily.'[26.]

It was truly the end of an era for Lyme Regis.

TEN
Legacy and Legend

WITH MARY ANNING'S DEATH Lyme Regis had lost one of its main tourist attractions. A decade later *The beauties of Lyme Regis, Charmouth, the Land-slip, and their vicinities*, a guidebook by Henry Rowland Brown lamented the economic effect of the demise of their celebrity fossilist:

'The decease of Miss Anning may be regarded even in a pecuniary point of view, as a great loss to the neighbourhood, as her presence attracted a large number of distinguished visitors, who able to appreciate her genius, were desirous of perambulating with her, those shores which she had made celebrated: delighting to listen to her interesting descriptions, and instructive conversation.'[1.]

There were other fossil dealers in Lyme, although none with the fame and reputation of Anning. Local directories for the 1840s also list as fossilists William Moore & Son on Broad Street, and Jonas Wishcombe in Charmouth. The Marder Brothers, pharmacist James Marder and his brother, surgeon Henry Marder, also dealt in fossils and ran a museum or fossil exhibit in Lyme. In his guidebook Rowland Brown noted that

'At Miss Anning's decease, her valuable collection was sold, but visitors, on application to Mr. J. Marder, can be favored by an inspection of several valuable and highly interesting specimens of the Ichthyosaurus Plesiosaurus, &c.'

This implies that the Marders acquired some of Anning's stock. In the advertisements at the back of Brown's guidebook, both James and Henry have separate notices of their fossil collection available for viewing, with some specimens for sale, in 'the large room at the Baths, Lyme Regis'. From the descriptions given in Henry's advertisement, their collection was substantial, including 'one skeleton of the Ichthyosaurus Platyodon nearly 30 feet long is the largest yet discovered.'[2.]

Nor was fossil-dealing restricted to Dorset. Another fossil shop which operated at about the same time as Anning's and also sold Lias marine reptiles lay far to the northeast along the outcrop of Jurassic rocks in the small Yorkshire fishing village of Whitby. There, a carpenter called Brown Marshall and his son were active from before 1823 to 1878 supplying ichthyosaur and plesiosaur specimens, many to the Yorkshire Philosophical Society. Tourism and the fossil trade were much less developed in Whitby than in Lyme, and although some of the Marshalls' discoveries were reported in the press, they never achieved the recognition, reputation and renown of the Annings. There were also shops selling fossils in Scarborough, Bridlington and Filey.[3.]

The business of fossil collecting and dealing in Lyme and Charmouth continued through the second half of the nineteenth century. On 15 March 1848, a year after Mary Anning's death, on the recommendation of George Roberts, William Moore wrote to Henry De la Beche, now well-established in London running his new Geological Survey and Museum of Economic Geology, to offer him a 'good specimen of Ichthyosaurus', five feet long, which his wife had found the previous week, for £15. Moore's business was, like the Annings', a family one, with his wife Sarah playing an active role.[4.]

From the 1850s, Moore, who also sold music, pianos and pictures, operated from a building called the Fossil Depot at the bottom of Broad Street, on the corner of Cobb Gate and Bridge Street opposite the Pilot Boat Inn. Often mistaken for Anning's shop in accounts of her life, it became a local landmark, featuring in photographs and postcards of Lyme in the late nineteenth and early twentieth centuries. It was also sketched by the Glaswegian architect and designer Charles Rennie Mackintosh when he visited Lyme in 1895. Photographs of the Fossil Depot often show a large whale shoulder blade to the left of the doorway. This bone is now in Lyme Regis Museum and although it has been suggested that it had previously been displayed from about 1833 outside Mary Anning's shop on Broad Street, there is no proof of this. By the early twentieth century, the Fossil Depot was run by a fishmonger, Sidney Curtis who sold fossil and fresh fish from the same counter. Visiting Lyme while researching his 1906 book, *Highways and Byways of Dorset*, the Dorset-born London surgeon Sir Frederick Treves noted: 'The old fossil shop ... is as curious a house as any in the town. Here can be purchased, at the same counter, fresh prawns or fossil ammonites, filleted soles or pieces of a saurian's backbone.' The Fossil Depot was demolished in 1913 along with other buildings on the south side of Bridge Street when the street and its narrow Buddle Bridge were widened.[5.]

In Charmouth, 1850s directories list John Hunter, variously as a fisherman, 'fossiler' and pleasure boat owner, while the 1861 census lists him as a 'dealer in fossils', a trade continued by his son, Isaac, who was still active in 1906 when he was visited by eighty-five members of the Geologists' Association on their Easter field excursion. Also collecting fossils in Charmouth was Samuel Clark, listed as a fossilist or fossil dealer in the census from 1851 to 1891, and his son, William.

Commercial fossil collecting and selling, in the tradition of Anning and her contemporaries, continues to this day in Charmouth and Lyme. Today's collectors work in very much the same way as Anning, searching out fossils after the storms and heavy rain, and then devoting many hours of patient preparation in their workshops to remove the sediment in which the fossils are embedded. They have a few techniques which were unavailable to Anning such as power and compressed air tools which speed up manual preparation, and the use of acids to dissolve fossil bone from the limestone. Like Anning two hundred years before them, they are saving new and exciting fossils from destruction by the sea and making them available for scientific examination and museum display.[6]

MARY ANNING'S BOOKS AND PAPERS

Soon after Anning's burial her family began disposing of her possessions. On 3 April 1847, less than a month after her death, the engineer and palaeontologist Proby Thomas Cautley, known for his work on the Ganges Canal in India but home in England for three years due to ill-health, visited Lyme and purchased from Anning's executors several of her books, including her copy of George Roberts' 1823 *The History of Lyme-Regis*, and John Phillips' 1841 *Figures and descriptions of the Palaeozoic fossils of Cornwall, Devon and West Somerset*.[7]

The book on Palaeozoic fossils may have been a gift from John Phillips as an indication of the style of a book he was planning on belemnites, about which he and Anning had corresponded through the 1830s. In 1836 Phillips advertised his *Figures and descriptions of British belemnites* which he hoped to publish by subscription in 1838. Despite a grant from the British Association for the Advancement of Science in 1841 to help him complete his research, a keen publisher in the form of John Murray, and encouragement from the newly-established Palaeontographical Society in 1848, it was another twenty years before it was published, in five parts between 1865 and 1870. In his monograph, Phillips acknowledges the belemnite specimens sent to him by Mary Anning.[8]

In the summer of 1885, the Earl of Enniskillen purchased a portfolio of manuscripts, drawings and other papers which had belonged to Mary

Anning from her nephew, Albert Anning. Enniskillen passed them straight on to Richard Owen at the British Museum suggesting he take whatever he liked then pass the remainder to several of his colleagues. By the time of Owen's death in 1892, what remained of these Anning papers had been subsumed into Owen's own huge collection of manuscripts and letters, all of which were stored in a cowshed at his former home in Richmond Park. Charles Davies Sherborn from the British Museum (Natural History) was invited by Owen's family to take on the enormous task of sorting them out as Owen's grandson was preparing a biography of his grandfather. Once that was published, Sherborn was given all of the papers which he then 'distributed to those interested all over the world,' amongst them some of Anning's. Fortunately, some do survive within the Owen Papers in the Library of the Natural History Museum.[9.]

One of the items Sherborn retained was Enniskillen's letter to Owen about his Anning purchase, and its enclosure of a letter from Albert Anning to Lord Enniskillen listing the papers, so we know at least some of the manuscripts and drawings that Mary Anning possessed. These include printed copies – reprints – of published papers such as the 1821 paper by De la Beche and Conybeare on ichthyosaurs and plesiosaurs, the 1835 paper by Sir Philip Egerton on *Ichthyosaurus,* and the catalogue from the 1820 auction of Colonel Birch's collection as well as Mary Anning's own hand-copied versions of several published papers. Amongst the prints and drawings she owned were various illustrations of fossils, mainly fish, ichthyosaurs and plesiosaurs, and several drawings and prints by Henry De la Beche probably given to her by him, including his cartoon of Buckland, *A Coprolitic Vision.* There were various notes, letters and poems and her Fourth Notebook, her commonplace book. The collection of papers purchased by Enniskillen probably represents only a small part of those Mary Anning possessed. Where are her first, second and third notebooks, for example, or the bulk of her correspondence? Several times she complains of the volume of correspondence she has to deal with, so there must have been a large number of letters to Anning, unless she disposed of them as she went along. Whatever was left after her death must have been dispersed, destroyed or lost sometime before 1885.[10.]

Anning also possessed a copy of J.S. Miller's 1821 book, *A Natural History of the Crinoidea*, inscribed to 'Miss Mary Anning'. It was included in the Earl of Enniskillen's 1885 purchase of the Anning papers but it is now lost, having last been seen in 1983 when it was listed for sale in the catalogue of a Bristol book dealer.[11.]

MARY'S RELICS

Apart from manuscript material, of which there may be, perhaps, a hundred or so items scattered through the world in well over a dozen different collections, and her fossils, there are few surviving artefacts that have a proven association with Mary Anning. We don't have her bonnet, basket or boots, for example. Poor people tended to pass such things on to their relatives or to wear them beyond all further use, but for some of her contemporaries, we do have such items: Adam Sedgwick's boots are in the Sedgwick Museum in Cambridge, as are some of his geological hammers, along with those of William Buckland and William Daniel Conybeare; hammers belonging to Buckland, Murchison and Lyell are in the Geological Society Archives; and a tabletop made of Scottish coprolites that once graced Buckland's drawing room is now in Lyme Regis Museum.[12.]

LYME REGIS MUSEUM AND MARY ANNING

Lyme Regis is fortunate to have had some notable local historians. The most prominent in the first half of the nineteenth century was George Roberts who was born in Lyme about five years after Mary Anning and knew her for much of her adult life. A century after Roberts, Cyril Wanklyn took on the mantle of Lyme historian, and over the last fifty years local history research continued with the work of Jo Draper and the novelist John Fowles, who despite his fame as author of *The French Lieutenant's Woman*, for many years served as the museum's honorary curator. The museum remains a centre for research into the history of the town and maintains a very active group of volunteer researchers. All have added to our knowledge of Mary Anning and the town where she lived.[13.]

The museum, however, came into existence long after the death of Mary Anning. It was founded by Lyme town councillor and mayor, Thomas Embray Davenport Philpot, a great-nephew of the Philpot sisters. The building was completed in 1901 as part of a redevelopment of Cockmoile Square begun with a remodelling of the Guildhall for Queen Victoria's Jubilee in 1887. The museum site purchased by Philpot that year had been previously been occupied by several houses. It was only realised when an 1824 map of Lyme was acquired by the museum in 1985 that one of those houses had been lived in by the Annings from about 1808 to 1826. Nothing of it now remains, although its footprint is followed at least in part by the museum building.

Quite what Thomas Philpot intended the museum to display is unclear. His great-aunts' fossil collection had been given to Oxford University twenty years earlier, so once completed, the museum building lay largely empty until

1921 when the first exhibits were installed.[14.]

Although the museum dates from a century after Mary Anning's birth, it has, over the years, acquired some items which may – or may not – have an association with her. Perhaps the object with the most convincing link to Mary Anning is the small metal disc stamped with her name found by a metal detectorist in 2014, and with whose discovery this book begins.[15.]

One of the most celebrated items is an object which until recently was considered to be Mary Anning's 'fossil-extractor', a makeshift hammer, made, legend has it, for her by her father. A metal-sheathed wooden pick with a chisel-edge at one end and a pointed pick at the other, and a broken wooden shaft, it bears no resemblance at all to the hammer that Anning holds in her famous portraits – but geologists do tend to own several different hammers and this may have been made when she was a child. To a geologist, however, the 'fossil-extractor' seems too flimsy a tool for breaking open rocks. In 2012, Richard Bull, a retired geologist and now a researcher at the museum, decided to take a closer look at 'Mary Anning's fossil-extractor'. He discovered that it had been found in the museum as recently as 1993 and had been catalogued then simply as 'a pick'. Of its provenance there was no record, nor did there seem to be any indication that it had any link with Mary Anning. But at some date after 1993, there emerged the belief that it had once been hers. Sadly, that is not the case. With the help of colleagues in other museums, Richard has now identified Mary Anning's fossil-extractor as the broken handle of a British Army Mark I entrenching tool of 1882.[16.]

Another object in the museum that was reputedly used by Mary Anning to extract fossils is a small pointing trowel. Although trowels are usually more tools of the archaeologist rather than the geologist, they can be useful to prise apart thin layers of shale or to lever out fossil bones from softer rocks. As with the 'fossil-extractor', there is no known provenance to this object that would connect it with Mary Anning.

The museum does, however, have an Anning tool: not Mary's, but perhaps Joseph's. This is a large cabinet-maker's try square on which is inscribed the name 'ANNING'. It was acquired by the museum in a purchase of tools from a firm of joiners and undertakers in Lyme. The manufacturers, Maw & Staley, were in business between 1823 and 1835, which is precisely the period during which Joseph Anning was working as an upholsterer and cabinet-maker, so it is quite possible that this try square belonged to him. With his carpentry skills, Joseph most probably helped his sister with the construction of wooden frames into which she set her marine reptile discoveries in plaster. This try square, if Joseph's, could well have been used to ensure the square corners of such frames.

PORTRAITS OF MARY

One of the few objects which does have a good provenance and association with Mary Anning is her portrait which today resides in the Natural History Museum in London, with a near-contemporary copy in the Geological Society. The image we have today of Mary Anning is of a dark-haired woman in a wide-brimmed bonnet and long green cloak, with a basket over her arm and clutching a hammer. This is based solely upon these much-reproduced portraits. Both of these pictures probably portray a true likeness of Anning in her forties. One was painted from life, the other a copy, but by an artist who knew her.

In February 1842, according to a later-annotated copy of George Roberts' 1834 *History and Antiquities of the Borough of Lyme Regis and Charmouth*, 'Mr Grey took a portrait for the Exhibition at the Royal Academy. This has become the property of Mr Joseph Anning, Miss M. Anning's brother.' The artist may have been William Gray, a sculptor, landscape and portrait painter, then based in London, who exhibited at the Royal Academy intermittently between 1841 and 1857. Although Roberts states that the picture was intended for exhibition at the Royal Academy, it is not listed in any of the the Academy's exhibition catalogues for that period, so was probably never shown. The unsigned portrait, painted in oil on board, shows Mary Anning a few months before her forty-third birthday and five years before she died. We know nothing of any connection between Gray and Anning, or why it was painted, other than that Gray intended it for the Royal Academy's exhibition. It probably it came into the possession of her brother after Mary's death in 1847 which suggests that it may have been owned by her. Perhaps it hung in her shop or in her home.[17]

After Joseph Anning died in 1849, the picture passed to his widow, Amelia, and through the family to her grand-daughter, Annette, who presented it to the British Museum (Natural History) in 1935.[18]

In this picture, Mary stands on rather ill-defined foreshore rocks, her left hand pointing down to her little dog who lies curled on the rocks next to an ichthyosaur skull which appears, faintly, in the lower right corner of the picture. In the background can be seen Golden Cap and the coast east of Charmouth.

In some 1920s correspondence, Mary Anning's great-niece referred to the dog in this painting as 'Tray'. It is not clear if Tray was a replacement for the dog which was killed in 1833 when the cliff fell on him, or is the one that died and was added to the painting by Gray, not from life, but from Anning's own sketch of her long-dead pet drawn by her in a similar pose in about 1824.[19]

Another, near identical portrait, but this time in pastel, hangs in the foyer of the Geological Society's apartments at Burlington House on Piccadilly and was acquired by the Society some sixty years before the Natural History Museum received theirs. Unlike the Natural History Museum portrait, this one is signed and dated, 'B.J.M. Donne pinxit 1850', the date telling us that it was drawn three years after Mary Anning's death. Benjamin John Merefield Donne was brought up in Crewkerne in Somerset close to its boundary with Dorset. In 1841 Donne came to Lyme as a pupil at George Roberts' school in Broad Street. As the school was across the road from Mary Anning's fossil shop, Donne would certainly have known Anning and been familiar with her appearance. The portrait is clearly a copy of the oil painting by Gray so Donne must have been given access to this earlier picture by Amelia Anning, Joseph having died the year before. Donne would have been a young man of only nineteen years old when it was painted. In the 1920s, Cyril Wanklyn tracked down the artist, then in his nineties and living in Exmouth, but he had no recollection of the painting. Wanklyn suggests that the portrait was a commission by the donors of the Anning memorial window in St Michael's Church, and its date would certainly support that idea.[20]

Drawn in a different medium, Donne's portrait is crisper, clearer and brighter than Gray's. Mary Anning appears slimmer and less stocky, despite her voluminous dress and cloak, her eyes a bright blue and looking younger than her years. A loose block of rock containing an ammonite has been added lying on the ground between Anning and her dog, and in the background, not only are Golden Cap and Thorncombe Beacon better defined, but the foreshore rock ledges where many of her specimens were collected are apparent. The spiral shell of an ammonite is also visible in the rocks behind Anning, below her basket. This picture was presented to the Geological Society in 1875 by William Willoughby Cole, 3rd Earl of Enniskillen, and had presumably remained in Lyme since it was painted, as Enniskillen had arranged for James Marder, the Lyme chemist, to send it to the Society.[21]

Intriguingly, Roberts' later annotations in his 1834 book suggest that two other portraits were painted of Mary Anning. One a watercolour, painted about 1838 by a 'private lady ... for Miss Penning' and a good likeness, and the other 'perfectly unlike', painted by 'Miss Wyse, sister of Mr Wyse M.P. for Waterford ... for a Museum in Ireland'. This was probably Harriet Wyse who also painted a portrait of her brother. Neither of these pictures of Mary Anning has yet been traced.[22]

Mary Anning may – or may not – also feature in the foreground of an 1825 lithograph of Lyme by Charmouth artist Thomas Carter Galpin. Standing on the shoreline gazing out to Lyme Bay is a slim young lady in a bonnet with a hammer over her right shoulder. She seems not to notice the weight of what appears to be a sizeable and no doubt heavy ammonite in her left hand. Is this Anning? Possibly, but it could equally represent a fashionable female fossilising visitor to Lyme.[23.]

Another image which has been identified as Mary Anning since the 1930s is a watercolour sketch showing a rather round figure with a hammer, and wearing a cloak and top hat, perhaps as protective headgear. Attributed to Anning's friend geologist and artist Henry De la Beche, it has been widely reproduced in biographies and papers about Anning and for a time served as the logo of Lyme Regis Museum. Unfortunately, recent research suggests that it is not a picture of Mary Anning, nor is it by De la Beche; nor was it drawn in Lyme Regis. In fact, it is a drawing of William Buckland researching the effects of glaciers in Snowdonia in North Wales in 1841, sketched by geologist and engineer Thomas Sopwith. [24.]

Mary Anning's life overlapped only briefly with the invention in 1839 of photography. The earliest photographs known of Lyme date from the 1850s and there was no photographer based in Lyme before 1861.[25.]

Recently, it has been suggested that a photograph taken in 1843 of a woman, dressed in a manner not dissimilar to that of Anning in the paintings, and a man apparently examining a rock outcrop at Chudleigh in Devon, is of Mary Anning and Henry De la Beche. Both figures have their backs to the camera, so their faces cannot be seen. Entitled *The Geologists,* the picture was taken by the pioneer photographer William Henry Fox Talbot and resides in the collection of the National Science and Media Museum in Bradford. While a photograph of Mary Anning would be a remarkable discovery (there are several known photographs of De la Beche who had close family connections with some pioneer photographers in Swansea), it is extremely unlikely that *The Geologists* is a photograph of Anning. The woman's fashionable costume is that of a middle class lady, not the plain clothes of a working woman like Anning; Chudleigh is situated between Exeter and Newton Abbot, so some considerable distance from Lyme and not known to have been visited by her; and the fact that the man is bare-headed suggests that the couple are either married or related. More likely the photograph shows two members of Fox Talbot's family posing at a well-known Devon beauty spot.[26.]

THE LEGEND AND THE MYTH

Thanks to the contemporary press coverage Mary Anning's discoveries received, and to the writings of George Roberts and the popular educational author Maria Hack, Anning was literally a legend in her own lifetime. But it has been said that 'soon after her death, she was forgotten ... an uneducated little girl, with a quick mind and an accurate eye, played a key role in setting the course of the 19th century geologic revolution. Then – we simply forgot about her' or that 'she was largely forgotten by the time Charles Darwin published *On the Origin of Species* in 1859'. Anning is often described as a 'forgotten' figure of palaeontology, but this assumption is not borne out by the many books, newspapers and magazine articles in which she was mentioned, many published long after her death. Even Gideon Mantell, who seemed so unimpressed by Anning and her shop when he visited in 1832, lauds her in his 1851 book, *Petrifactions and their teachings*. Adam Sedgwick in 1869 recalled her as 'a collector of early celebrity' and Richard Owen lamented her loss in his 1870 review of Lias pterosaur fossils. In 1877, Cotswold geologist Robert F. Tomes described a new species of Lias coral which he had found at Charmouth and named it *Tricycloseris anningi*.[27.]

She was commemorated again in 1927 by Scottish-born South African physician and palaeontologist Robert Broom when he named a new genus of mammal-like reptile after her. Broom was one of the first to publicly state a feeling that she had not received the acclaim she deserved:

'I propose to call it *Anningia megalops* in honour of Miss Mary Anning, of Lyme Regis. Though many of the finest specimens of fossil reptiles in the British Museum were discovered by Miss Anning, and these specimens formed the basis of much of the work of Home, Conybeare, De la Beche, Hawkins, and Owen, and thus helped to give Britain its high position in the history of vertebrate palaeontology, I have long felt that the part played by Miss Anning has not been fully appreciated as one of the worlds greatest fossil hunters and as a pioneer. Although it is about a hundred years since she did most of her work, one may perhaps still be allowed to lay a stone on the cairn of this most remarkable woman.'[28.]

Four years later, Leslie Reginald Cox of the British Museum (Natural History), who had spent childhood holidays with his grandfather in Charmouth, described a new genus of fossil bivalve from Dorset, *Anningia*, 'named after Miss Mary Anning, the famous collector of Lyme Regis fossils'. Cox, however, was unaware that this name had already been used by Broom, so in 1958 he was obliged to change the name to *Anningella*. This is still a valid name and several species of *Anningella* are known from the Dorset Lias.[29.]

More fossils have been named after her in more recent years. In 1974, a palaeontologist at University College London named a Dorset fossil ostracod – a tiny crustacean – *Cytherelloidea anningi*. It took two hundred years, however, for Anning's name to be attached to the marine reptiles for which she was most famed. In 2012, palaeontologists described a new genus of plesiosaur from Lyme, naming it *Anningasaura* and in 2015, a new species of *Ichthyosaurus* was named after her.[30.]

From 1885 to 1900, the great and the good (and some of the bad) of British history were given a place in the sixty-three volumes of the *Dictionary of National Biography*. Only three per cent of the 29,104 entries were women; needless to say, Mary Anning was not amongst them. However, in 1901, the *DNB* publishers issued a *Supplement* containing 'a thousand articles, of which more than two hundred represent accidental omissions from the previously published volumes'. One of these was Mary Anning, whose four hundred word entry in the *Supplement* was written by Bernard Barham Woodward, Librarian of the British Museum (Natural History). It is not the most accurate account of her life and achievements, but her inclusion shows us that at least by 1901 she was recognised as a significant figure in British history by her entry in the *DNB*, which for a provincial, working-class woman was exceptional. By 1910 she even had her own entry in the *Encyclopaedia Britannica*.[31.]

Anning's memory lived on in the popular press, too, with articles in provincial newspapers and magazines from the 1860s onwards as well as in travel guidebooks to the West Country from the 1850s to the present day, including one in 1911 which christened her 'the Saint Georgina of Lyme Regis' for slaying the dragon-like 'dread pterodactyl'. Some were more accurate accounts than others, the more reliable being by those who knew her. Henry Rowland Brown knew Anning when he was a child and had access to Anning family papers for his 1857 and 1859 Lyme guidebooks, while an anonymous article published in *Chambers's Journal* in 1857 is now known to have been written by Frank Buckland, who knew Anning from visiting Lyme with his father, William.[32.]

Less reliable are an 1865 article in *All the Year Round*, a weekly literary magazine founded, owned and edited by Charles Dickens, and a small book for children published in 1925, both of which have introduced elements of myth into the Anning legend. The popular *All the Year Round* carried a mix of fiction and non-fiction articles contributed by authors like Elizabeth Gaskell, Anthony Trollope, and Wilkie Collins, as well as by Dickens himself. *Mary Anning, the fossil finder*, was, like many contributions, published anonymously, but it is often attributed to Dickens. It is, however, now thought to have been written by a Bath schoolmaster, the Reverend Henry Stuart Fagan. Publishing in Dickens' magazine brought the story of Mary Anning to a wide audience at the time. Unfortunately, the *All the Year Round* piece is, according to modern Anning scholars, 'a careless hack job ... which drew heavily, and often erroneously, on other writers'. It cannot be relied upon as a useful source.[33.]

Fagan's article was extensively copied and reprinted in provincial newspapers, including in a French translation. Four years later the story was rewritten as *The little fossil-gatherer* by Isabel Thorne in *Chatterbox*, a weekly magazine for older children. Although derived from Fagan's article, this may be the first telling of the Mary Anning story written specifically for children. The cover engraving of this issue shows a young Mary, hammer in hand, collecting ammonites which she passes to her father for their collecting-basket. This image was to be the start of a tradition of depicting Anning in print and picture as the child fossil-collector. This theme was picked up in an 1884 verse of what W.D. Lang has termed 'dreadful doggerel' by J.W. Preston:

> 'Miss Anning, as a child, ne'er passed
> A pin upon the ground;
> But picked it up, and so at last
> An Ichthyosaurus found.'[34.]

The Anning child tale was reworked in 1925 with the publication of a slim little book, *The Heroine of Lyme Regis: The Story of Mary Anning the Celebrated Geologist* by Harriet Anne Forde, predominantly an author of religious tracts, and published almost thirty years after her death. A semi-fictionalised account, it is couched as a morality tale of self-help and Victorian values. This book introduces the tale that Anning's dog, which Forde calls 'Fido', would sit guarding a newly-found fossil on the beach until its mistress returned; what evidence there is for this is unclear. Anning's story was further promulgated by its appearance in the hugely successful *Children's Encyclopaedia* which was translated and distributed widely around the world through the first half of the twentieth century.[35.]

Her childhood find of the ichthyosaur, along with the lightning strike, are the two most prominent strands of the Anning legend, a tale told and retold in over thirty children's books published since 1960, including some in Spanish and Japanese, and in many thousands of websites. However, as historian Hugh Torrens has long pointed out, with so much of the literature about Anning focusing on this childhood discovery, her adult work risks being neglected. This may go some way to explaining why she is so often considered 'forgotten', her find as a child overshadowing and detracting from her achievements as an adult, and reducing her significance and importance. It is no help to her cause that she has for so long been one step removed from her tangible legacy, her fossils, her name unrecorded by museums.[36.]

The progress of geology through the nineteenth and into the twentieth century also may have contributed towards Anning's recession into the mists of public consciousness. When she was active, geology – and palaeontology in particular – was the cutting edge field of scientific endeavour, and there was widespread interest and public enthusiasm for the new discoveries of strange fossils. This excitement continued in 1841-42 when Richard Owen recognised that the large land-living fossil reptiles could be grouped together as the Dinosauria, the dinosaurs, and with the construction of life-size dinosaur and other fossil animal models at the Crystal Palace in 1854. By the 1870s, however, the focus of fossil reptile, especially dinosaur, research had moved to the American West, and as the century closed and the new one began, discoveries in physics took scientific precedence. By 1949, W.D. Lang could write 'Mary Anning is now remembered mainly by geologists who are interested in West Dorset'. Geology's return to public prominence has come only since the 1960s with the development of a new understanding of how the earth operates – plate tectonics – and more recently with remarkable new feathered dinosaur fossil discoveries in China.[37.]

ANNING FOR ADULTS

No one can doubt the attraction in Anning's story. A poor, working class woman of limited education, in a small remote town, making remarkable discoveries which challenge the established orthodoxy of creation, supporting her widowed mother through selling fossils, developing an expertise that allows her to debate with learned professors, yet remaining impoverished while the gentlemen who dominate nineteenth century society exploit her finds to establish their own reputations in the new science of geology to which she cannot make a formal contribution, barred by her gender from membership of the Geological Society and from contributing to their debates. Add in a hint of unrequited love and you have rich material for novels such as those by Tracy Chevalier and Joan Thomas. Although not a story about Anning, Sarah Perry's best-selling 2016 novel, *The Essex Serpent*, has the principal character's passion for fossils stemming from her meeting an elderly gentleman who knew Anning.[38.]

In addition to the written word, Anning's story has been seen on stage several times. A successful play, *Fossil Woman*, ran in London in the 1990s and dealt with a series of episodes in Anning's life. More recently, *She sells sea shells*, an 'extraordinary story of genius, gender and dinosaurs' was performed in London and at the 2019 Edinburgh Festival Fringe. Other plays about Mary Anning include *The excavation of Mary Anning* and *Mary Anning – the mad woman of Lyme!*[39.]

Anning's story was told, perhaps for the first time on screen in the

41. 'Verteberries', isolated vertebrae from ichthyosaur backbones, commonly sold as souvenirs to visitors to Lyme and Charmouth from the late eighteenth century on.

42. *Ammonites obtusus* from Lyme illustrated by James Sowerby in his *Mineral Conchology of Great Britain*, vol ii, plate 167, 1817. Now called *Asteroceras obtusum*, the foreground specimen was from the collection of Elizabeth Philpot.

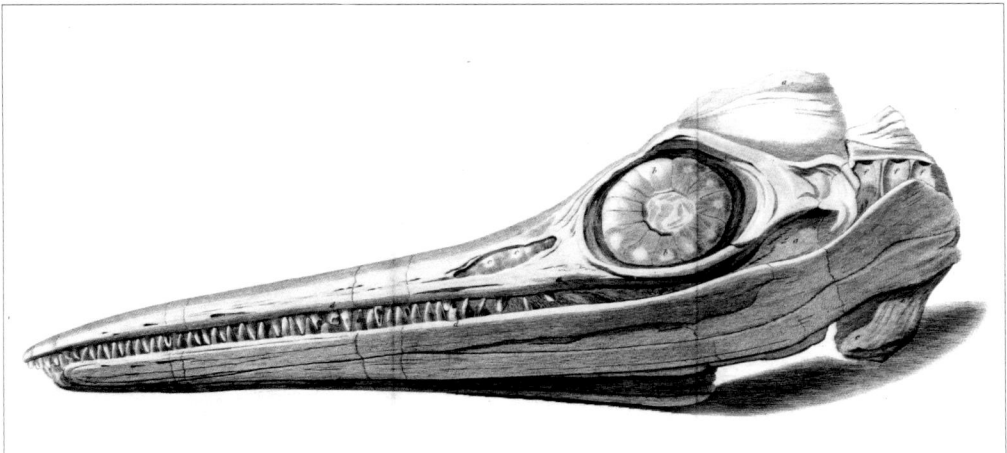

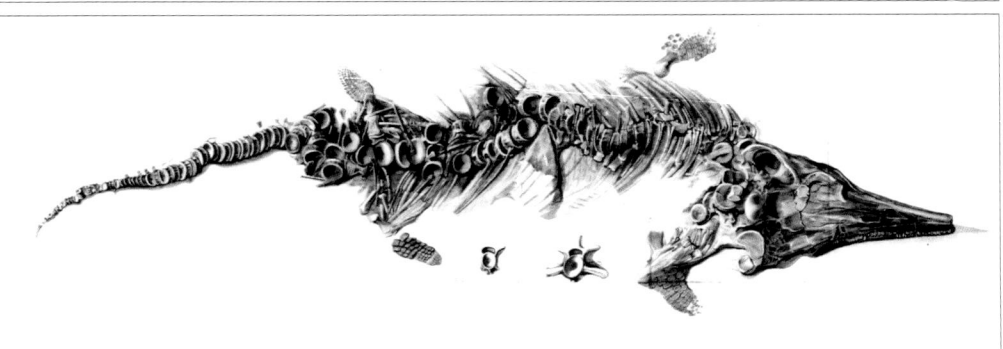

TOP 43. Ichthyosaur skull found by Joseph Anning in the autumn of 1811 between Lyme and Charmouth and now in the Natural History Museum, London.

CENTRE 44. The Anning ichthyosaur skull illustrated by Sir Everard Home in the *Philosophical Transactions of the Royal Society* in 1814.

BOTTOM 45. Colonel Birch's complete ichthyosaur skeleton, probably purchased from the Annings. Illustrated by Home in 1819.

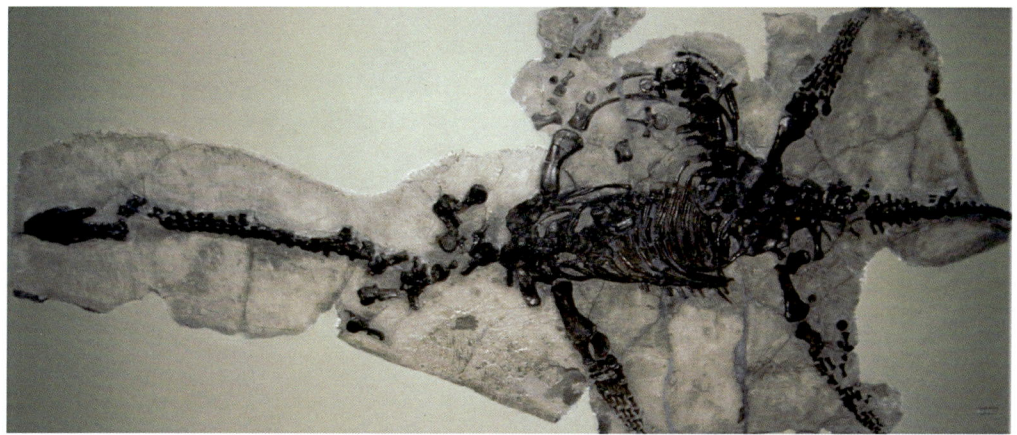

46. The first complete *Plesiosaurus*, discovered by Mary Anning in December 1823 and described by W.D. Conybeare at the Geological Society in 1824.

47. Mary Anning's 1823 complete *Plesiosaurus dolichodeirus*. *Transactions of the Geological Society*, v.1, plate 48, 1824.

48. W.D. Conybeare's sketch of Mary Anning's 1823 plesiosaur in his letter describing its discovery to De la Beche, 6 March 1824.

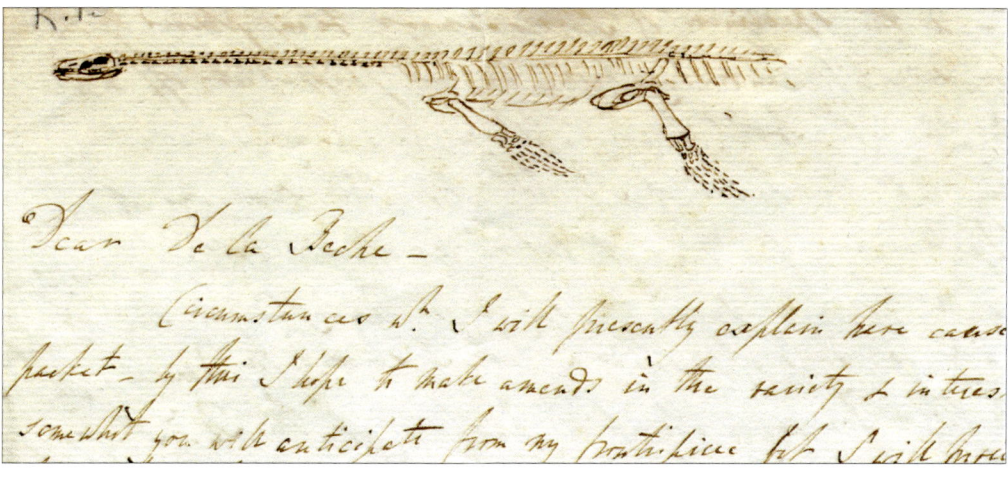

49. 'Pterodactyle' (*Dimorpodon macronyx*) partial skeleton found by Mary Anning in December 1828.

50. Rev. George Howman's 1829 reconstruction of Mary Anning's pterosaur.

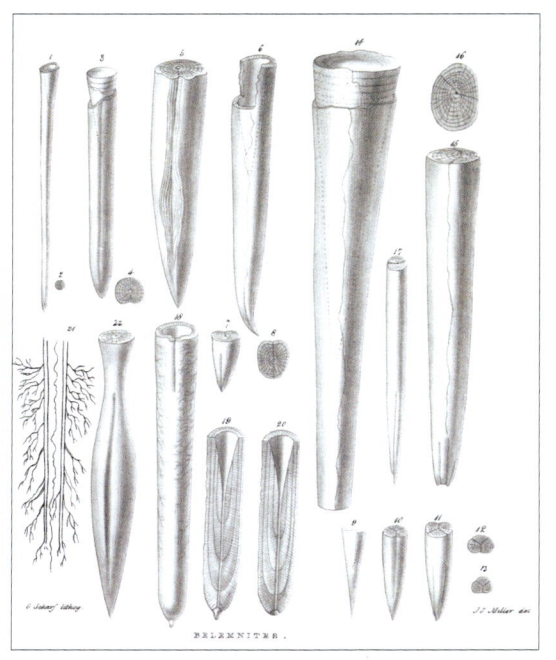

51. Belemnite fossils from J. S. Miller's 1826 paper Observations on belemnites published in the *Transactions of the Geological Society*.

52. Mary Anning's carefully hand-drawn copy of Miller's plate.

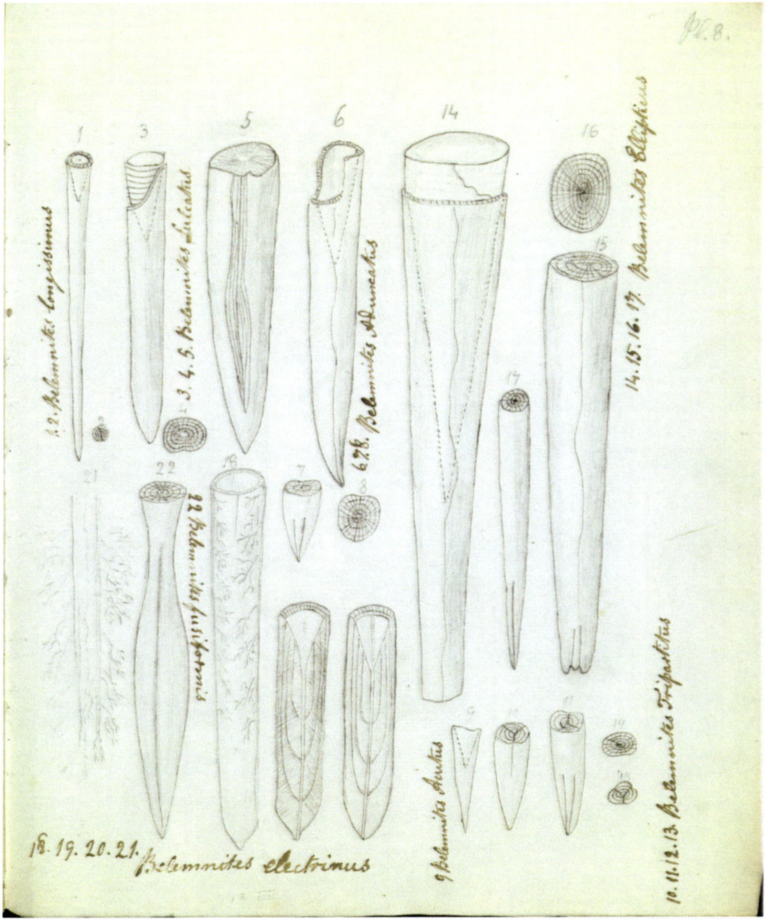

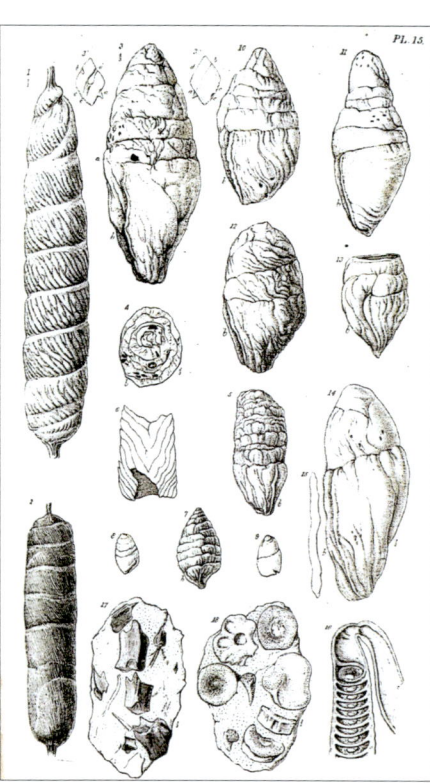

LEFT 53. 'Coprolites chiefly from the Lias of Lyme Regis' from William Buckland's 1836 *Bridgewater Treatise*.

BELOW 54. William Buckland's coprolite table top made about 1834 and containing fish coprolites from near Edinburgh.

OPPOSITE PAGE TOP 55. *A Coprolitic Vision*, a humorous sketch by Henry De la Beche c.1829 celebrating William Buckland's fascination with coprolites.

OPPOSITE PAGE CENTRE & BOTTOM 56, 57. Coprolite, Lias, Lyme Regis, showing spiral structure noted by Buckland, and containing black undigested fish scales. The lower photograph is an enlarged detail of the fish scales.

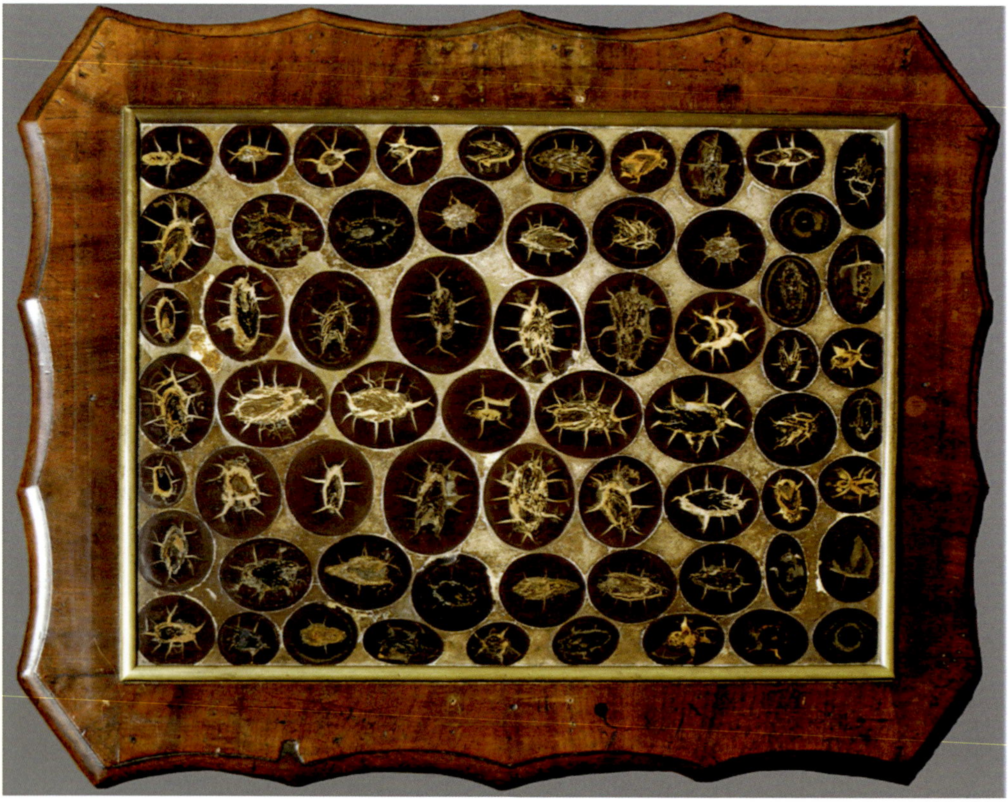

58, 59. *Duria Antiquior,* a watercolour of the Lias seas of Dorset by Henry De la Beche in 1829-30 showing Mary Anning's discoveries as living animals, the first ever reconstruction of its kind and (below) as updated by Lyme artist Richard Bizley in 2007, a modern interpretation of early Jurassic life in Dorset.

60. Lithograph of *Duria Antiquior* published by De la Beche and sold to raise money for Mary Anning in 1830.

61. *Awful Changes*, an ichthyosaurian professor lecturing about a human skull by Henry De la Beche, 1830. Published as a lithograph, this is probably the original artwork, given to Mary Anning by De la Beche.

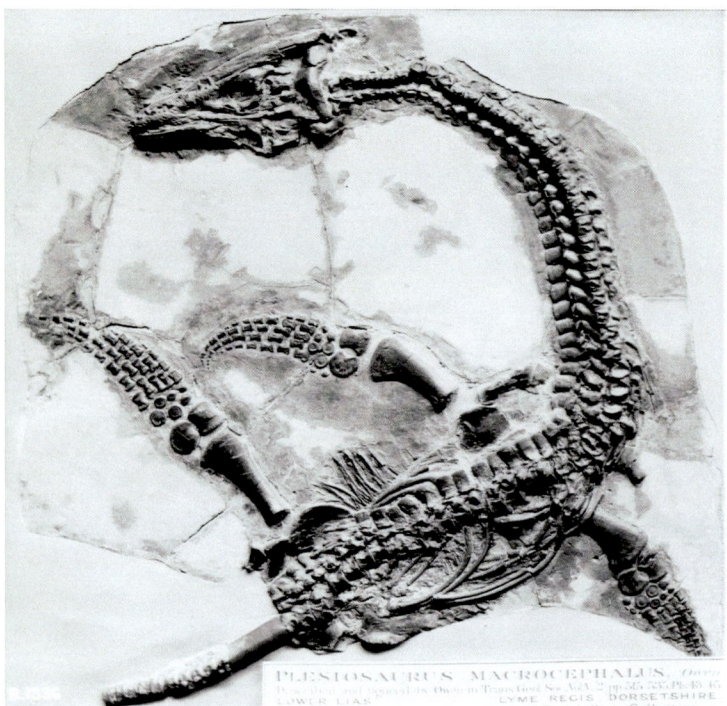

TOP 62.
Plesiosaurus macrocephalus, discovered by Mary Anning in 1830 and named by William Buckland in 1836.

CENTRE 63.
The fossil fish *Squaloraja* discovered by Mary Anning in December 1829 and recognised by her as a new species.

BELOW 64.
'*Ichthyosaurus chiropolyostinus*' collected by Thomas Hawkins between Lyme and Charmouth in 1833 after Mary Anning told him it would fall to pieces. Plate 7 of Hawkins' 1834 book *Memoirs of Ichthyosauri and Plesiosauri*.

65. Frontispiece by J. S. Templeton in Thomas Hawkins' 1834 book *Memoirs of Ichthyosauri and Plesiosauri*.

66. The fossil collector Thomas Hawkins (1810-1889).

67. Frontispiece by John Martin in Thomas Hawkins' 1840 *The book of the great sea dragons*.

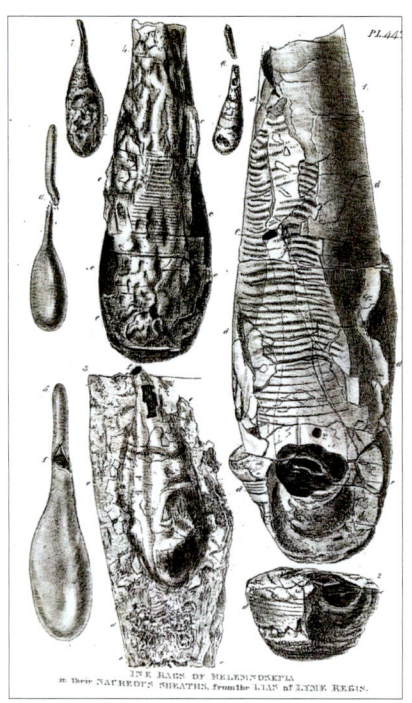

LEFT 68. Fossil sepia ink bags discovered by Mary Anning in 1828 (right) and Charlotte Murchison (top, centre). From William Buckland's 1836 *Bridgewater Treatise*.

ABOVE 69. Ichthyosaur skull drawn by Elizabeth Philpot using ink from fossil sepia in a letter to Mary Buckland, 9 December 1833.

BELOW LEFT 70. 'Pentacrinites briareus from the Lias at Lyme Regis, Dorset. In the collection of Professor Sedgwick'. Now called *Pentacrinites fossilis*, this may be a specimen purchased by Sedgwick from Mary Anning. Plate 53 from Buckland's 1836 *Bridgewater Treatise*.

BELOW RIGHT 71. A new species of fossil crinoid discovered by Mary Anning near Bridport in June 1835 and drawn by Elizabeth Philpot in a letter to William Buckland.

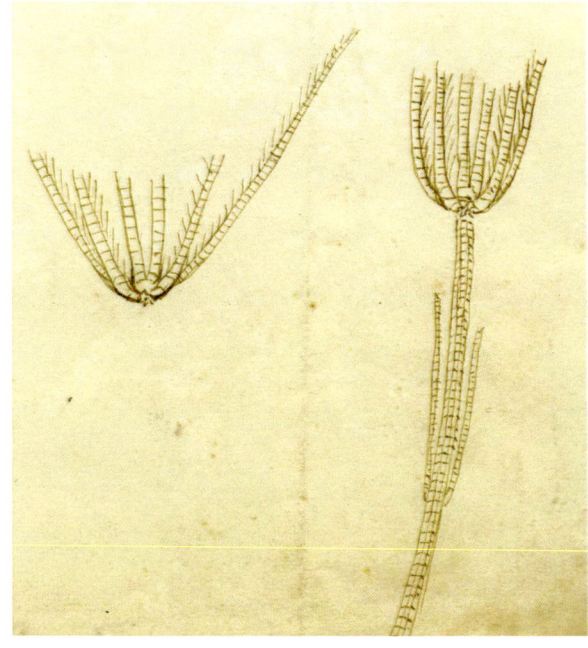

72. Fossil brittle star, *Palaeocoma egertoni*, collected by Mary Anning and purchased from her by the British Museum in 1840.

ABOVE RIGHT 73. *Dapedium punctatum* from the collection of the Philpot sisters. Louis Agassiz described it as 'the finest example of fossil fish I have ever seen'.

74. *Ichthyosaurus anningae* with a coprolite within its ribcage. Purchased from Mary Anning by the British Museum in the 1830s.

75. Ichthyosaur with preserved stomach contents found by Mary Anning in the early 1830s and illustrated in Buckland's 1836 *Bridgewater Treatise*. Now recognised as a specimen of *Ichthyosaurus anningae*, it is in the collection of Oxford University Museum of Natural History.

76. Large skull of *Temnodontosaurus platyodon* presented to Bath Literary and Scientific Institution by W.H. Eastwick in 1825 and probably purchased from Mary Anning.

77. The Bindon landslip of Christmas 1839 drawn by Exeter surveyor William Dawson. Mary Anning accompanied William and Mary Buckland and W.D. Conybeare when they went to examine it soon after the event.

78. St Michael's Church, Lyme Regis, where Mary Anning worshipped later in life.

79. Gravestone of Joseph and Mary Anning in St Michael's Churchyard. Probably erected by Amelia, Joseph's widow, there is no definitive proof that the stone marks Mary's grave.

80. Mary Anning Memorial Window, installed in 1850 in St Michael's Church by the Rev. F. Parry Hodges and members of the Geological Society.

81. Thin metal disc discovered at Church Cliffs in 2014. Stamped with Mary Anning's name and the date 1810 on one side, and Lyme Regis and her age of eleven on the other, it may have been made for Mary by her father.

82. Wax seal on a letter from Mary Anning to William Buckland, 21 December 1830, made with a stamp engraved with her name, 'Mary'.

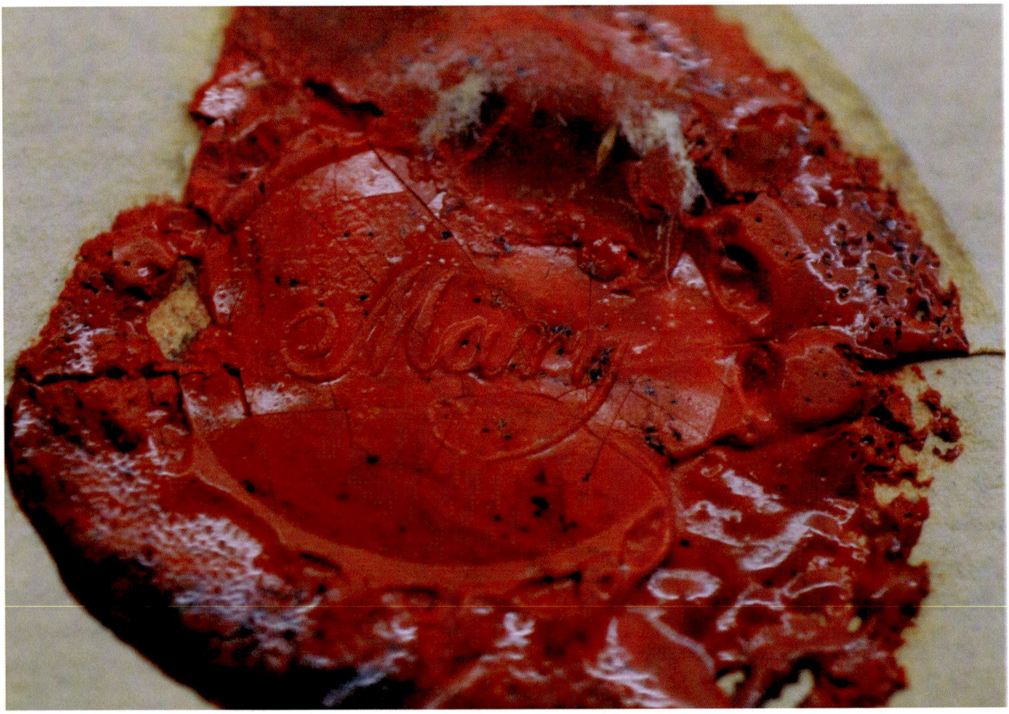

LEGACY AND LEGEND

early 1970s, in play called *The Crocodile* filmed on location in Lyme for BBC Schools television. Lyme was also the location for the filming of the 1981 film *The French Lieutenant's Woman*. Although Mary Anning plays no part in John Fowles' novel, set two decades after her death, one of the main characters' interests was fossil-collecting and her fossil shop merits a mention. Anning and Lyme both feature in two films scheduled for release in 2020 or 2021, *Ammonite*, already mentioned, and *Mary Anning and the dinosaur hunters*.[40.]

DID SHE SELL SEA SHELLS ON THE SEA SHORE?

Perhaps the most pervasive myth of Mary Anning is that she is the subject of, or inspiration for, the tongue-twister rhyme *She sells sea shells on the sea shore*, and that this was originally a song written about her in 1908 by a music hall lyricist called Terry Sullivan and put to music by Harry Gifford. It was certainly performed in the first decade of the twentieth century by the Lancashire comic, singer and actor Wilkie Bard who introduced tongue-twisting songs into his music hall and pantomime act.[41.]

In 2017, the American folklorist Stephen Winick of the Library of Congress's American Folklife Center delved into the history of *She sells sea shells* and found that while Sullivan and Gifford turned it into a song, it did not originate with them. Winick found a version dating from 1855 published as an elocution exercise by Alexander Melville Bell, a lecturer on speech and elocution at the University of Edinburgh. Bell has it as just four words, 'She sells sea shells', but through the latter part of the nineteenth century it evolved into an amusing tongue-twister with the addition of lines, including 'He sells sea shells', and 'Shall Susan sell sea shells'. In no version is there any mention of Mary Anning or her discoveries. She sold fossils, not sea shells, and the fossils for which she was best known were not fossil shells but reptiles. If nothing else, the tongue-twister has proved a useful hook for teachers introducing children to Anning's life and work.[42.]

RECOGNISING MARY ANNING TODAY

In 1999, to mark the bicentenary of Anning's birth, a large international audience of palaeontologists, historians and fossil collectors gathered in Lyme for a major three-day conference which addressed the theme of *Mary Anning and her times: the discovery of British palaeontology, 1820-1850*. The packed programme reviewed our knowledge of Anning, the society in which she lived, her position as a woman in a male-dominated science and society, her associates and contemporaries, the fossils she found, and her impact on the development of vertebrate palaeontology. Plesiosaur

161

palaeontologist, Mike Taylor, noted that while Anning's 'gentlemanly patrons are now almost forgotten, except by historians ... Anning's specimens and story are increasingly giving her a permanent identity in research and public display.' [43.]

In 2002, Anning was commemorated by the Palaeontological Association when it established a new prize, the Mary Anning Award, to recognise outstanding contributions to palaeontology made by amateurs. The conditions of the award specify that nominees must not be professionally employed in palaeontology, which, amusingly, would exclude Mary Anning, the professional fossil dealer, herself. In 2010, the Royal Society considered her one of the ten most influential women in the history of British science, and in 2018, we almost had her portrait on a new banknote. The educational value of her story has long been recognised informally but that recognition is now official as Anning now features in the National Curriculum Key Stages 1 and 2 in England.[44.]

In Lyme, a plaque on the town's museum commemorates the site of the first Anning fossil shop and on Broad Street a small bronze plaque – confusingly on the wall of the adjacent building on the downhill side – records: 'In a house on this site Mary Anning (1799-1847) died March 9th 1847', a very understated marker for such a key place in the history of palaeontology. A residential street is named after her, Anning Road. St Michael's Church has the memorial window and the churchyard has her gravestone with its remembrances of small fossils and flowers left by visitors. Her spirit hovers over the Lyme Regis Fossil Festival which attracts palaeontologists both amateur and professional to the town. Some feel that a more substantial and tangible memorial is required, so in 2019 a campaign was launched for a statue of Anning to be erected in Lyme.[45.]

* * * * *

In 1992 Stephen Jay Gould described Anning as 'probably the most important unsung (or inadequately sung) collecting force in the history of paleontology'. He was probably correct in saying the geological community's singing of her praises has been inadequate, and that given her achievements, Anning is not as famous as she ought to be. But she has never really been forgotten. Her name and work have been remembered throughout the century and a half since her death. About forty books have been written about her, the majority for younger readers. All tell the story, with varying degrees of accuracy, in a way which appeals to their readership and play an important role in bringing Mary Anning to the attention of a new and young audience, potentially inspiring others – girls in particular – to follow in her footsteps and develop an interest in science. She features in many works on the history of palaeontology as well as in numerous academic papers in the field of the history of geology. In fact, more has been written about Mary Anning than any other British geologist apart from Charles Darwin. She is the lead character in novels, plays and films. An internet search of her name throws up about a million results. It is testimony to her achievements that two centuries on we retain our fascination with her.[46]

In recent decades there has been a revival of interest in recognising the significance of Mary Anning's contribution to palaeontology. This stems from a combination of three main strands of study. Firstly, a renewed interest by palaeontologists in fossil marine reptiles which has led to a new understanding of their anatomy, relationships and way of life – a new Golden Age, it has been called. This has brought modern scientists into direct contact with Anning's specimens and a need to know more about what she did and how she did it. Secondly, a growth of interest by geologists in the history of their science with the formation of organisations like the Geological Society's History of Geology Group. And thirdly by an increased awareness of the largely unrecognised contribution made by women to science generally and to geology in particular.[47]

In her day, the 'celebrated Miss Mary Anning,' 'this persevering and successful *collectress* of extraneous fossils,' was a major attraction bringing visitors to Lyme. Today she continues to make a contribution to the economy of the town as tourists, students, and professional palaeontologists flock to follow in her footsteps. Visit Lyme on any almost any day of the year and you will find someone, hammer in hand, picking over the rocks at Black Ven or Church Cliffs searching for fossils, just as the 'fossil woman of Lyme Regis' did two hundred ago. But her fame has outgrown the quest for fossils. It is now also as a woman, the daughter of an age in which women had no voice, that Mary Anning is finally being heard – and justifiably honoured.[48]

Mary Anning's Fossils

We will never know the whereabouts today of all of the fossils collected by Mary Anning. There are no records of what must have been the day-to-day sales of thousands of small fossils to Lyme's visitors, specimens such as ammonites, bivalves and crinoids or perhaps single reptile vertebrae and teeth, or fragments of ichthyosaur jaw. Such few records as do exist note the sale or purchase mainly of some of the larger, more complete (and now famous) vertebrate specimens or those mentioned in the contemporary literature as having been collected by Anning. These were purchased either directly by museums, or by wealthy collectors whose collections subsequently may have been donated to or purchased by museums. In this process the name of the original collector is often lost, the institution's ledgers recording the beneficence of the donors. When the British Museum purchased the two collections of Thomas Hawkins in 1834 and 1840 and that of the 3rd Earl of Enniskillen in 1882-83, the museum acquired specimens which both of these collectors had obtained from Anning, although they were not necessarily identified as such. This will also be the case with other collections from that period later acquired by other museums. Where specimens may have been bought by an institution directly from Anning, even then her name might not have been recorded. It was not common in the nineteenth century for the vendor's name to be noted; it was merely a commercial transaction, and museum ledgers might simply list the specimen as 'purchased' with no other details. As a result, specimens collected by Mary Anning probably lie unrecognised in many of the older, long-established museums around Britain and in some museums overseas.

However, Anning specimens can be found in some museum collections, although not necessarily on display. The most notable collections of Anning material are in the Natural History Museum in London, Oxford University Museum of Natural History, the Sedgwick Museum, Cambridge, and the Muséum National d'Histoire Naturelle in Paris.

NATURAL HISTORY MUSEUM, LONDON

Around thirty Anning specimens have been identified in the Natural History Museum, and many are on display. Most are fossil reptiles and include some of the Annings' most famous specimens, such as their first ichthyosaur, *Temnodontosaurus platyodon*; the complete plesiosaur found in 1823, *Plesiosaurus dolichodeirus* and another of the same species found in 1829; the 1830 plesiosaur, *Plesiosaurus macrocephalus*; and the 1828 pterosaur, *Dimorphodon macronyx*. The museum's curators have also identified fossil squid, starfish, fish and coprolites and others as having been collected by Anning. Some of the specimens in the collections of Thomas Hawkins and Enniskillen are recognised as having come from Anning and it is likely that more lie within the Hawkins, Enniskillen and Egerton collections.[2]

The loss of associated information on even famous historical specimens can be compounded by the passage of time. A good example is the large ichthyosaur skull found in 1811-1812 and bought by the British Museum at Bullock's sale in 1819. In 1945 W.D. Lang wrote that the 'first Ichthyosaur found by Mary Anning, that which she obtained in 1811, when she was eleven or twelve years old ... has disappeared since the dispersal of the saurians in Bullock's Museum in 1820'. The skull of this specimen was eventually recognised in the collections of the Natural History Museum – where Lang himself had been Keeper – in 1969. Although part of the skeleton survives attached to the skull, much of the remainder, noted at the time of its discovery as 17 feet (5 m) in length, has not been located.[1]

OXFORD UNIVERSITY MUSEUM OF NATURAL HISTORY

In Oxford, specimens attributable to Mary Anning are surprisingly few, given her close association with William Buckland. Eight specimens are recognised as having come from Anning. These include the fossil squid *Phragmoteuthis*; a belemnite ink-sac from Watchet on the Somerset coast; a plesiosaur paddle; and a coprolite collected and presented by Anning. The star Anning specimen at Oxford is a skeleton of a young ichthyosaur which was presented to Buckland by Lord Cole. This was one of the first ichthyosaurs found with preserved stomach contents and was illustrated in Buckland's 1836 *Bridgewater Treatise*. Originally identified as *Ichthyosaurus communis* and collected by Anning sometime before 1836, it has recently been recognised as a specimen of *Ichthyosaurus anningae*, a species named after Mary Anning and described for the first time in 2015.[3]

Oxford University Museum of Natural History also houses the collection of the Philpot sisters which was presented to the museum in 1880. After the death of Elizabeth Philpot in 1857, their home, Morley Cottage, was occupied by their nephew, John Philpot, who moved from London with his wife Elizabeth Mary. Following John Philpot's death his widow presented the sisters' fossil collection to Oxford University in memory of her husband. Some, if not many, specimens in the Philpot collection would have been found when Elizabeth Philpot went fossil-hunting with Mary Anning. The collection includes the tail of the fossil fish *Squaloraja*, which was found separately from the rest of the skeleton discovered by Anning and which went to the Bristol Institution.[4]

SEDGWICK MUSEUM, CAMBRIDGE

Adam Sedgwick at Cambridge University was an active purchaser of specimens from Mary Anning from the 1820s to the early 1840s. In 1984, Sedgwick Museum curator, the late David Price, searched the Sedgwick's collections and records for these specimens and found that there was no mention of Anning in the museum's catalogues or on any of the labels. This was a surprise as he had just become aware of the letters from Anning to Sedgwick in Cambridge University Library and a book of 1869 with a preface written by Sedgwick:

'In the year 1819 I procured a few specimens of the Ichthyosaur from the Lias of Somersetshire ... and in the year following some additions were made to the Reptiles of the Lias, during an excursion along the coast of Dorsetshire. In several subsequent years valuable specimens were purchased from Mary Anneing of Lyme Regis, a collector of early celebrity. Among them were two very good specimens of the Ichthyosaur; and a very beautiful Pentacrinite.'[5]

With the assistance of a museum volunteer, Price found not only the letters but also details of Sedgwick's transactions with Anning in his field notebooks and account books. This led Price to identify a number of specimens which were most likely those purchased from Anning, although in many cases it was impossible to be certain. Others are unlikely ever to be identified as some of the records are vague, such as an 1821 receipt signed by Mary Anning for 'various fossils 12s 0d' and 'part of Ichthyosaurus £2 10s 0d'. One specimen, however, was unequivocally from Anning, an ichthyosaur skeleton which she had, very conveniently, sketched in her letter to Sedgwick on 4 May 1843. In addition to this ichthyosaur, Price identified five others which were probably Anning specimens as they matched descriptions in the letters.[6]

One of these specimens was described in 2013 by two ichthyosaur researchers, Judy Massare and Dean Lomax, who identified it as an example of *Ichthyosaurus breviceps*. Interestingly, on close examination, they noticed that part of the backbone was not only upside down compared with the rest of the skeleton, but it seemed to belong to a different species. At some point the specimen had been modified with the addition of these bones, perhaps to improve its appearance for display. It seems unlikely that Anning would have done this, considering how critical she was of Hawkins' habit of altering and adding to his specimens. Her knowledge of ichthyosaur skeletal anatomy was such that she would never have been so clumsy as to insert the bones upside down.[7]

BATH ROYAL LITERARY AND SCIENTIFIC INSTITUTION
NATIONAL MUSEUM OF WALES, CARDIFF

Sometimes, Anning specimens have a complex history which takes some unravelling. The Department of Geology of the National Museum of Wales in Cardiff houses a large collection of ichthyosaurs built up in the mid nineteenth century by the Bath collector Charles Moore. Moore's collection, mainly from the Lias rocks of Somerset, formed a spectacular display in the Bath Royal Literary and Scientific Institution in the 1860s. In 1963 the ichthyosaurs were lent to a researcher, the late Dr Bob Appleby of what was then University College of South Wales, Cardiff (now Cardiff University), and stored in the National Museum. Amongst them is a large skull of *Temnodontosaurus platyodon* which Anning expert Hugh Torrens has recognised as one presented to the Bath Literary and Scientific Institution in 1825 by William Henry Eastwick, the engineer and agent of the Kennet and Avon Canal Company. At some stage Eastwick's specimen had become conflated with the Moore collection. This *Temnodontosaurus platyodon* skull comes, not from Somerset, but from Lyme Regis and it is probable that Eastwick purchased the specimen from Mary Anning. The skull, which is almost six feet long, is one of the largest known.[8]

The Moore collection at Cardiff also includes another Lyme Regis specimen, *Ichthyosaurus communis,* which was presented by John Templeman who had purchased it from Mary Anning sometime before 1828. Templeman lived in Bath but had a summer home in Lyme Regis, Temple House on the lower part of Broad Street, and made several purchases from Mary Anning which he donated to the Bath Institution, including fine examples of the fish *Dapedium politum* in 1828 and the crinoid *Pentacrinus briareus* in 1837.[9]

SOMERSET COUNTY MUSEUM, TAUNTON

Also in Somerset, in the collection of the County Museum in Taunton there is an ammonite from Lyme Regis, thought to have been collected by Anning in 1838. It was presented to the Somerset Archaeological and Natural History Society in 1935 as part of a larger collection of fossils.[10]

SAFFRON WALDEN MUSEUM, ESSEX

Within the geology collections of Saffron Walden Museum should lie a crinoid specimen, probably *Pentacrinites*, which was purchased from Anning by local banker Wyatt George Gibson in 1839. Gibson's brother, Jebez was involved in its formation and had responsibility for the geology collections when the museum opened in 1835.[11.]

BRISTOL CITY MUSEUM

A remarkable survivor is the rear part of another large ichthyosaur skull with a huge eye socket in the collections of Bristol City Museum, perhaps only the second skull of this species to be discovered by the Annings. It was found near the low water mark at Lyme in 1813 and was purchased by the Bristol collector James Johnson, who had, by then, been collecting in the Lyme-Charmouth area for about twenty years. Johnson paid ten guineas for the specimen and a further two guineas to have it transported to Bristol. A drawing of the specimen was made by Johnson's fellow Bristolian and fossil collector George Cumberland in 1814. When Johnson's collection was auctioned in 1845 following his death, this partial skull was bought by John Naish Sanders for £27 10s, having outbid the British Museum, for the Bristol Institution. The Institution had been established in 1823 and had amongst its early supporters William Conybeare and Henry De la Beche. Its collections contained many fossils purchased from Mary Anning, including an ichthyosaur described as the finest found in Europe and acquired in 1823, and the fish *Squaloraja* which she found in 1829. The Institution's collections eventually passed to Bristol City Museum. Sadly, this magnificent collection was destroyed on 24-25 November 1940 when a fire bomb landed on the museum building during the Bristol Blitz. In 1987, Bristol Museum's Curator of Geology, Peter Crowther, discovered a large partial ichthyosaur skull with a large eye socket in one of the museum's stores. With Hugh Torrens, he showed that it matched the drawing made by Cumberland in 1814. Somehow, this Anning specimen survived the bombing; it may have been stored elsewhere when the bomb struck.[12.]

GEOLOGICAL SOCIETY OF LONDON

Another Anning-Johnson specimen survives, at least in part, in the apartments of the Geological Society in Burlington House on Piccadilly in central London. Again, this is the skull of the large ichthyosaur *Temnodontosaurus platyodon*. Sometime between 1814 and 1817, the Annings discovered the back part of a skull and sold it to James Johnson. Both this specimen and

Johnson's other large *platyodon* skull from 1813 were examined by De la Beche in April 1819 when he visited Johnson in Bristol during the course of his research on ichthyosaurs.

About ten years after the discovery of the back part of this skull, Mary Anning discovered the front half. It was bought by De la Beche who presented it to the Geological Society in April 1827. For almost twenty years, the back of the skull was in Bristol and the front in London. At the 1845 auction of Johnson's collection, the Bristol part was purchased by a group of six Fellows of the Geological Society who presented it to the Society so that the two halves were finally reunited and the skull was now complete. So far, so good, but it was to transpire that this reunification was not to be permanent.

In 1874, the Geological Society moved from its premises at Somerset House to new accommodation at Burlington House where there was space to display the geological collections built up since its foundation in 1807. By the start of the twentieth century, pressure on space led to the Society's governing Council reconsidering its need to maintain its own collection. In 1911, the collections were removed and divided between the Geological Survey's Museum of Practical Geology, then nearby on Jermyn Street, which took the British specimens, and the British Museum (Natural History) which received the foreign material. The Society decided to retain a small number of specimens for display at Burlington House, amongst which was this Anning-De la Beche-Johnson skull. By the late twentieth century its link to Anning had been lost and the skull languished in the Society's basement. Sadly, by 1998 the rear part of the skull, purchased at the Johnson sale in 1845, had disappeared, presumably discarded by builders clearing the basement. The front part, presented by De la Beche in 1827, survives and today its historical connections are realised and celebrated.[13.]

In November 1827, the year in which De la Beche presented the ichthyosaur skull, William Henry Fitton, the Society's President, instructed the clerk, Samuel Taylor junior, to arrange for a locksmith to break open a locked table drawer in the Society's library. When it was opened, they found inside a couple of keys, a fossil fish specimen, and several boxes. One small, empty tin box was labelled 'Miss Anning'. What this box might once have contained is a mystery; it may have held a fossil sent by Anning to the Society, or may have contained something that was intended for her. Sadly, the box was not kept. It does, however, again show that Anning was known at the Geological Society.[14.]

ROYAL COLLEGE OF SURGEONS

Bristol Museum was not the only one to lose Anning marine reptile fossils to bombing during the Second World War. On the night of 10 May 1941, the museum of the Royal College of Surgeons in Lincoln's Inn Fields in London was badly damaged by a bomb and the fossil reptile collection was almost completely destroyed. This important and historic collection contained fossil marine reptiles from the late eighteenth century, predating any Anning discoveries, acquired by the surgeon John Hunter, although their true nature was not recognised at that time. Additional material was added in the nineteenth century by Everard Home and by Richard Owen who, in 1854, published a catalogue listing about four hundred marine reptile fossils in the collection.

Many of those from Lyme Regis are recorded simply as having been 'Purchased', but others were donated by Conybeare, Buckland, Home or the surgeon James Luke. The purchased specimens include some used by Home in his papers on what he called *Proteosaurus*, and amongst them is the complete ichthyosaur bought to Home's attention by De la Beche in 1818 and bought by the College from Colonel Birch's sale in 1820. This is, or rather was, most likely an Anning specimen and it is probable that many of the other purchased specimens from Lyme also came from the Annings. Of the four hundred marine reptile fossils in the collection in 1854, just four survived the bombing, none of them from Lyme.[15.]

BRITISH GEOLOGICAL SURVEY, KEYWORTH

The British Geological Survey was founded as the Ordnance Geological Survey in 1835 by Henry De la Beche. A museum to house the specimens collected by the Survey staff was soon established in Craig's Court off Whitehall in London and this later expanded into a new purpose-built Museum of Practical Geology in Jermyn Street in 1851. When the Geological Society disposed of its museum collections in 1911 the British specimens were transferred to the Museum of Practical Geology.

Only two specimens have been identified in the Survey's collections as having an association with Mary Anning. One is a fossil fish, *Hybodus delabechei*, first described in 1839 and named after De la Beche by Edward Charlesworth. Found by Anning, it was purchased from her by Edmund Thomas Higgins and at some stage made its way into the Survey's collection.

The other is also a fossil fish from Lyme Regis, *Dapedium politum*, which was collected by Charlotte Murchison, probably while out fossil-hunting with Mary Anning on her visit in 1825. It may have came into the Survey's collections sometime during the tenure of Roderick Murchison as Director of the Survey.[16.]

NATIONAL MUSEUMS OF NORTHERN IRELAND, BELFAST

An eight foot (2.4 m) long specimen of *Ichthyosaurus communis* from Lyme Regis on display in the museum in Belfast was, like one of the Geological Survey's specimens, originally in the collection of Edmund Thomas Higgins and was donated by him to the Yorkshire Museum in 1848. The *Annual Report of the Council of the Yorkshire Philosophical Society* records a 'Beautiful specimen of an Ichthyosaurus discovered at Lyme Regis by the Donor', but probably he had purchased it from Mary Anning. In the 1970s, it was acquired by the Ulster Museum through exchange from the Yorkshire Museum.[17.]

ANNING ABROAD

Anning specimens also found their way overseas to France, Germany and the United States, and probably elsewhere. In 1820 several ichthyosaur specimens were purchased at the sale of Colonel Birch's collection for Georges Cuvier at the Muséum National d'Histoire Naturelle in Paris. They include a large, incomplete skull of *Ichthyosaurus communis* bought for fourteen guineas. Some of these specimens were illustrated in Cuvier's 1824 book on fossil bones, *Recherches sur les ossemens fossiles*. Also in the Paris collection is an almost complete plesiosaur skeleton from Lyme. Although missing part of its neck and its head, this was probably only the second partially complete specimen discovered. It was found in the rocks to the west of the Cobb not by Mary Anning but by retired Royal Navy captain Henry Waring, probably in January 1824 or perhaps in December 1823, so only a matter of weeks after the discovery of the first complete specimen. Anning bought Waring's specimen from him before selling it on, in June 1824 to French geologist Constant Prévost who was in Lyme buying specimens for Cuvier.

Other specimens of fossil marine reptiles from Lyme were also being sent to Cuvier by Buckland and Conybeare who probably collected them while out fossil-hunting with Mary Anning, so there may be more Anning material awaiting discovery in the Paris collection.[18.]

In Germany, the Naturkunde-Museum in Coburg in Bavaria, has a collection of fossils which belonged to Queen Victoria's husband, Prince Albert of Sachsen-Coburg and Gotha. Some, from the Lower Jurassic rocks of Lyme Regis, may have been collected by Mary Anning. Albert had wide interests in science and technology and was a Fellow of the Geological Society.[19.]

New York Lyceum of Natural History contained specimens purchased from Mary Anning in October 1827 by G.W. Featherstonhaugh but these did not survive a fire which destroyed the entire Lyceum collection on 21 May 1866.

In Philadelphia, at the Academy of Natural Sciences, two probable Anning specimens may survive, but are unrecognised and perhaps unrecognisable. In 1847, Thomas Bellerby Wilson presented four fossil marine reptile skeletons to the Academy. Two ichthyosaurs were from Somerset, and the third ichthyosaur and a plesiosaur were from Lyme and were probably bought from Anning. Wilson acquired many British fossils and other natural history specimens through his brother Edward Wilson of Tenby in South Wales. Like his brother in Philadelphia, Edward had wide interests in geology and natural history and was a Fellow of the Geological Society. He seems to have been actively acquiring specimens to send to Philadelphia. Over time, any original labels and documentation have been lost, as so often happened, so an Anning affiliation can be difficult to prove.[20.]

Who's Who

AGASSIZ, JEAN LOUIS RODOLPHE (1807-1873), Swiss palaeontologist specialising in fossil fish.
ANNING, AMELIA, née Reader (c.1797-1858), Joseph Anning's wife, Mary Anning's sister-in-law.
ANNING, ANNE, née Flood (c.1733-1812), of Sidbury, Devon. Mary Anning's paternal grandmother.
ANNING, JOSEPH (1796-1849), Mary Anning's elder brother, an upholsterer but who also maintained an interest in the family fossil business until at least 1832.
ANNING, MARY (Molly) née Moores or Moors (1763-1842) of Blandford, Dorset, Mary's mother.
*ANNING, MARY (1799-1847), of Lyme, fossil collector and dealer.
ANNING, RICHARD (c.1766-1810), born in Sidbury, Devon. Mary's father.
ANNING, WILLIAM (c.1730-1799), of Sidbury, Devon. Mary Anning's paternal grandfather.
AVELINE, WILLIAM HUDDLE (1771-1837), Henry De la Beche's step-father in Lyme Regis.
BELL, FRANCES AUGUSTA (1809-1825), a young invalid visitor to Lyme befriended by Mary Anning.
*BENETT, ETHELDRED (1775-1845), Wiltshire fossil collector.
BIRCH, LT.-COL. THOMAS JAMES (1768-1829), of Lincolnshire. Fossil collector, purchased specimens from the Annings. Auctioned his collection in May 1820, the proceeds going to aid the Annings.
*BUCKINGHAM, RICHARD TEMPLE NUGENT BRYDGES CHANDOS GRENVILLE (1776-1839), 1st Duke of, purchased Mary Anning's first complete plesiosaur in 1824.
*BUCKLAND MARY née Morland (1797-1857), illustrator, fossil and mineral collector, wife of William Buckland.
*BUCKLAND, REV. WILLIAM (1784-1856), Axminster-born Oxford geologist.
BULL, RICHARD, geologist and local historian, researcher at Lyme Regis Museum.
*BULLOCK, WILLIAM (1773-1849), naturalist and antiquarian, owner of the

Egyptian Hall, Piccadilly where the first Anning ichthyosaur skull was displayed until 1819.
CARPENTER, DR THOMAS COULSON (c.1778-1833), Lyme apothecary and physician; fossil collector.
COLE, WILLIAM WILLOUGHBY (1807-1886), Third Earl of Enniskillen, fossil collector, mainly of fossil fish; visited Lyme to collect with Mary Anning. Purchased her *Plesiosaurus* of December 1830.
CONGREVE, ANN (1758-1823) Lyme fossil collector who lent ichthyosaur fossils to De la Beche and Conybeare.
*CONYBEARE, REV. WILLIAM DANIEL (1787-1857), Bristol (and later Axminster)-based geologist who published on Anning's ichthyosaurs and plesiosaur discoveries.
CROOK, REVEREND JOHN (c.1766-1830), Charmouth's Congregationalist minister who baptised Mary Anning in Lyme; also a fossil collector.
CROOKSHANKS, JOHN (c.1738-1802), London coal merchant and fossil collector, based in Lyme from about 1798 until 1802 when he committed suicide at Gun Cliff.
CUVIER, GEORGES (1769-1832), Europe's leading comparative anatomist based at Muséum National d'Histoire Naturelle, Paris; initially cast doubt on authenticity of Anning's 1823 *Plesiosaurus*.
*DE LA BECHE, HENRY THOMAS (1796-1855), geologist, Lyme resident 1812-1830, friend of Mary Anning.
DE LUC, JEAN-ANDRÉ (1727-1817), Swiss geologist, visited Lyme in 1805, met Richard Anning.
DOLLIN, HARRIET (*died* 1892), fossil dealer in Lyme after Mary Anning.
*EGERTON, SIR PHILIP DE MALPAS GREY- (1806-1881), fossil collector, mainly of fossil fish; visited Lyme to collect with Mary Anning.
*FEATHERSTONHAUGH, GEORGE WILLIAM (1780-1866), Anglo-American geologist who visited Anning in 1827, purchasing specimens from her.
GLEED, REV. JOHN (1785-1870), Independent minister in Lyme Regis 1818-28, and fossil collector.
GOODHUE, REV. TOM W., US-based Anning scholar and author of several books and papers on her.
*HAWKINS, THOMAS (1810-1889) of Sharpham Park, near Glastonbury, Somerset, fossil collector.
HENLEY, HENRY HOSTE (1766-1833) of Sandringham Hall, Norfolk and Lord of the Manor of Colway. Purchased the Anning's 'first' ichthyosaur.
HODGES, REV. THOMAS (c.1783-1847), Curate at Charmouth (1818-1827), friend of Mary Anning and Adam Sedgwick.
*HOME, SIR EVERARD (1756-1832), comparative anatomist at the Royal College of Surgeons who published the first papers on ichthyosaurs.

JOHNSON, JAMES (c.1764-1844), fossil collector of Hotwells, Bristol who collected at Lyme and Charmouth from about 1789.
KENNAWAY, SARAH, née Johnson (c.1774-1855), fossil collector living in Charmouth, friend and supporter of Mary Anning.
*KONIG, CHARLES (1774-1851), Keeper of Natural History, British Museum
LANG, DR WILLIAM DICKSON (1878-1966), Keeper of Geology, British Museum (Natural History) 1928-38; retired to Charmouth, studied the local strata, and published much on Mary Anning.
*LYELL, CHARLES (1797-1875), geologist, visited Mary Anning in Lyme in 1824.
*MANTELL, DR GIDEON ALGERNON (1790-1852), Sussex doctor and fossil collector, visited Lyme in 1832.
MARDER, HENRY ((c.1818-1869)) surgeon, and James (1824-1888), chemist in Lyme, also ran fossil collecting and selling business.
*MILLER, JOHANN SAMUEL (1779-1830). Curator of the Bristol Institution; purchased fossils from Anning.
MOORE, WILLIAM (c.1795-1867), Lyme fossil and music dealer 1839-67.
MOORS, MARY née Mitchell, of Durweston near Blandford. Mary's maternal grandmother.
MOORS, ROBERT, of Durweston near Blandford. Mary's maternal grandfather.
*MURCHISON, CHARLOTTE née Hugonin (1788-1869), wife of R.I. Murchison, friend and correspondent of Mary Anning, hosting her visit to London in 1829.
*MURCHISON, RODERICK IMPEY (1792-1871), geologist, first visited Lyme in 1825 and met Mary Anning.
*OWEN, RICHARD (1804-1892), comparative anatomist at the Royal College of Surgeons, and later at British Museum.
PARRY HODGES, REV. THOMAS FREDERIC AMELIUS (1801-1880), vicar of St Michael's, Lyme, 1833-1880, officiated at funerals of Mary and Joseph Anning, and installed a stained glass memorial window to Anning in 1850.
PHILPOT, ELIZABETH (1780-1857), fossil collector resident in Lyme, along with her sisters MARY (1777-1838) and MARGARET (1786-1845).
PINNEY, ANNA MARIA (1812-1861), sister of William Pinney (1806-1898), MP for Lyme 1832-1842; confidante of Mary Anning.
PRÉVOST, CONSTANT (1787-1856), French geologist, visited Lyme in 1824 and purchased specimens from Mary Anning.
ROBERTS, GEORGE (1803-1860), Lyme schoolmaster and local historian.
*SEDGWICK, REV. ADAM (1785-1873), Woodwardian Professor of Geology, Cambridge University, 1818-1873. Purchased specimens from Mary Anning.
SOUTH, JOHN (c.1778-1831), fossil collector based in Lyme from about 1798 to at least 1813.
*SOWERBY, GEORGE BRETTINGHAM (1788-1854), London naturalist and dealer, Mary Anning's London agent.

STOCK, CHARLOTTE née Shapland (c.1779-1861) Lyme resident who employed the young Mary Anning to run errands and who gave her her first book on geology.

TAYLOR, DR MICHAEL A., palaeontologist and historian of geology who has published much on Mary Anning.

TEMPLEMAN, JOHN (c.1757-1848), philanthropist of Bath, had summer home on Broad Street, Lyme Regis, purchased specimens from Mary Anning.

TORRENS, PROF. HUGH S., geologist and historian of geology, the leading expert on Mary Anning.

WANKLYN, CYRIL (1864-1943) Lyme local historian, author of *Lyme Regis A retrospect*.

WISHCOMBE, JONAS (c.1787-1859), fossil collector and seller, Charmouth.

* Entry in Harrison, B. & Matthew, H.C.G. (eds), 2004. *The Oxford Dictionary of National Biography*. Oxford: Oxford University Press.

Notes, Sources & Further Reading

INTRODUCTION

1. For more on George Roberts (1803-1860), see Powell, C., 2018. *Lyme Regis Monographs*. FeedARead.com Publishing, 11-62; much about Lyme and Anning can be found in Roberts, G., 1823. *The History of Lyme Regis, Dorset, from the Earliest Periods to the Present Day*. Sherborne: Langdon & Harker; Roberts, G., 1830. *A guide descriptive of the beauties of Lyme Regis, being a sketch of the situation, salubrity, and picturesque scenery; with an account of environs, and a description of the great storm, in 1824*. Lyme Regis: Ham & Landray; Baldwin & Cradock; and Roberts, G., 1834. *The History and Antiquities of the Borough of Lyme Regis and Charmouth*. London: Samuel Bagster.

AUTHOR'S NOTE

1. For a useful discussion on the prices of fossils and the difficulties of such monetary comparisons see Taylor, M.A. & Torrens, H.S., 1987. Saleswoman to a new science: Mary Anning and the fossil fish *Squaloraja* from the Lias of Lyme Regis. *Proceedings of the Dorset Natural History and Archaeological Society*, v.108, 143-146; and Rolfe, W.D.I., Milner, A.C. & Hay, F.G., 1988. The price of fossils. *In* Crowther, P.R. & Wimbledon, W.A. (eds) *Special Papers in Palaeontology*, no.40, 139-171.

1. 'OLD WONDERS AND NEW IMPROVEMENTS'

1. Thomas Hollis (1720-1774). Much on the history of Lyme can be found in Roberts, G., 1823. *The History of Lyme Regis, Dorset, from the Earliest Periods to the Present Day*. Sherborne: Langdon & Harker; Roberts, G., 1834. *The History and Antiquities of the Borough of Lyme Regis and Charmouth*. London: Samuel Bagster; Brown, H.R., 1857. *The beauties of Lyme Regis, Charmouth and the landslip, and their vicinities; topographically and historically considered*. Lyme Regis: Daniel Dunster; Wanklyn, C., 1927. *Lyme Regis A retrospect*. Second edition. London: Hatchards; Fowles, J., 1982. *A short history of Lyme*

Regis. Wimborne: The Dovecote Press; and Draper, J., 2004. *Mary Anning's town – Lyme Regis*. Lyme Regis: Lyme Regis Museum.

2. Richard Russell (1687-1759). For more on the development of coastal resorts at this time see Walton, J.K., 1983. *The English seaside resort: a social history 1750-1914*. Leicester: Leicester University Press; and Allen, L., 2016. *The Georgian seaside. The English resorts before the railway age*. Louise Allen.

3. Fowles, J., 1982. *A short history of Lyme Regis*. Wimborne: The Dovecote Press, 26. Thomas Fane (1701-1771), 8th Earl of Westmorland; his second son, Henry Fane (1739-1802) was Member of Parliament for Lyme Regis from 1772 to 1802.

4. Henry Dinham Chard (1781-1812). For more on the maritime history of Lyme see Lacey, P., 2011. *Ebb & Flow. The story of maritime Lyme Regis*. Wimborne Minster: The Dovecote Press.

5. Wanklyn, C., 1927. *Lyme Regis A retrospect*. Second edition. London: Hatchards, 104-105.

6. Feltham, J., 1813. *A guide to all the watering and sea-bathing places, for 1813. With a description of The Lakes; a sketch of a tour in Wales; and itineraries*. London: Longman, Hurst, Rees, Orme, and Brown, 283-284.

7. Harriette Wilson (1786-1845). Her description of Lyme in 1806 is from *The memoirs of Harriet Wilson, written by herself*, first published in 1825 in London by J.J. Stockdale.

8. Roberts, G., 1823. *The History of Lyme Regis, Dorset, from the Earliest Periods to the Present Day*. Sherborne: Langdon & Harker, 136.

9. Jane Austen (1775-1817). *Persuasion* was Austen's last completed novel and was published six months after her death. The description of Lyme can be found in Chapter 11.

10. Jane Austen's letter mentioning Richard Anning was written on 14 September 1804 to her sister Cassandra and can be found in R.W. Chapman, 1952. *Jane Austen's letters to her sister Cassandra and others*. Second edition. London: Oxford University Press. Many of Austen's letters were destroyed by her sister, so any other references to Lyme have been lost. See also Austen-Leigh, W. & Austen-Leigh, R.A., 1913. *Jane Austen: her life and letters. A family record*. London, Smith, Elder & Co.; Honan, P., 1987. *Jane Austen: her life*. London: Weidenfeld & Nicolson; and Lane, M., *Jane Austen and Lyme Regis*. Chewton: The Jane Austen Society.

11. Roberts, G., 1834. *The History and Antiquities of the Borough of Lyme Regis and Charmouth*. London: Samuel Bagster, 179.

12. Taking its name from the Jura Mountains of France and Switzerland, the term 'Jurassic' was first used in 1799 as 'Jura-Kalkstein' (Jura Limestone) by the Prussian naturalist Alexander von Humboldt (1769-1859) and as 'terrains jurassiques' by the French mineralogist Alexandre Brongniart (1770-1847) in

1829. Lyme historian George Roberts defines Lias as 'a provincial term for, and western mode of pronouncing *layers* ... The local term *lias* was applied to the calcareous [limestone] beds; the thick argillaceous deposits [mudstones], now distinguished as the upper lias, was called *blue marl*.' Roberts, G., 1839. *An etymological and explanatory dictionary of the terms and language of geology designed for the early student, and those who have not made great progress in that science.* London: Longman, Orme, Brown, Green, & Longmans, 97-98. This dialect origin was disputed by the great Jurassic expert William Joscelyn Arkell (1904-1958) in his 1933 book *The Jurassic System in Great Britain* (Oxford: The Clarendon Press, 12). He supported the idea of palaeontologist Sidney Savory Buckman (1860-1929) that its source is the Gaelic (or Brythonic) word, *leac*, meaning a flat stone and seen in place names such as Lechlade and Leckhampton. Another possible origin for the term is from old French 'liois', a hard compact limestone, according to the *Oxford English Dictionary* and quoted by Arkell, W.J. & Tomkeieff, S.I., 1953. *English Rock Terms chiefly as used by miners and quarrymen.* London: Oxford University Press. For details of Jurassic rocks around Britain see Taylor, P.D. (ed.), 1995. *Field geology of the British Jurassic.* London: The Geological Society.

13. A great deal has been published on the geology of Dorset and its coast. Good, readable summaries are: Ensom, P., 1998. *Discover Dorset. Geology.* Wimborne: The Dovecote Press; Edmonds, R.,1999. *Discover Dorset. Fossils.* Wimborne: The Dovecote Press; Hart, M., 2009. *Dorset and East Devon landscape and geology.* Marlborough: Crowood Press; and Brunsden, D. (ed.) 2003. *A walk through time. The official guide to the Jurassic Coast. Dorset and East Devon's World Heritage Coast.* Wareham: Coastal Publishing. For more detailed geological information, see Cope, J.C.W. 2016. *Geology of the Dorset coast.* Geologists' Association Guide No. 22. 2nd edition. London: The Geologists' Association; Barton, C.M., Woods, M.A., Bristow, C.R., Newell, A.J., Westhead, R.K., Evans. D.J., Kirby, G.A. & Warrington, G., 2011. *Geology of south Dorset and southeast Devon and its World Heritage Coast. Special Memoir for 1:50 000 geological sheets 328 Dorchester, 341/342 West Fleet and Weymouth and 342/343 Swanage, and parts of sheets 326/340 Sidmouth, 327 Bridport, 329 Bournemouth and 339 Newton Abbot.* Keyworth, Nottingham: British Geological Survey; and Lord, A.R. & Davis, P.G., 2010. (eds), 2010. *Fossils from the Lower Lias of the Dorset coast.* Palaeontological Association Field Guide to Fossils: Number 13. London: The Palaeontological Association.

14. Mary Anning has been described by some authors as a discoverer of dinosaurs. As far as we know, she never found a dinosaur. The fossils for which Mary Anning is best known, ichthyosaurs and plesiosaurs, were marine reptiles, not land-living dinosaurs. Pterosaurs, another fossil group of which Anning found the first identified in England, were flying reptiles. Ichthyosaurs,

plesiosaurs, pterosaurs and dinosaurs were contemporaries, living during the Triassic, Jurassic and Cretaceous periods, from, roughly, 240 million years ago to about 66 million years ago, although some groups became extinct earlier than others. The first dinosaur bones from the Lias of Dorset were found in the Black Ven Marls, part of the Charmouth Mudstone Formation, on Black Ven in 1858 by James Harrison of Charmouth and described as a new species of plant-eating dinosaur named *Scelidosaurus harrisonii* by Richard Owen in 1861. See Lang, W.D., 1947. James Harrison of Charmouth, geologist (1819-1864). *Proceedings of the Dorset Natural History and Archaeological Society*, v.68, 103-118. Fossil insects in the Lias were first discovered in the 1950s by Charmouth fossil collector James Frederick Jackson (1894-1966).

15. Carus, C.G., 1846. *The King of Saxony's journey through England and Scotland in the year 1844. Translated by S.C. Davidson*. London: Chapman & Hall.

16. Letters from Mary Buckland and William Buckland to Elizabeth Philpot, 1 January 1834. Private collection.

17. 'X', 1892, *More Memories of Old Lyme*. Lyme Regis: J.S. Turner.

18. An excellent account of the evolution of our understanding of time is Wyse Jackson, P., 2006. *The Chronologers' Quest. Episodes in the Search for the Age of the Earth*. Cambridge: Cambridge University Press. The Jurassic Period extends from 201 million to about 145 million years ago. The Lower Jurassic (Lias) rocks around Lyme Regis on which Mary Anning mainly worked are between 201 million and 183 million years old. These and other dates for the whole of the geological column can be found on the *International Chronostratigraphic Chart* v.2020/03 published by the International Commission on Stratigraphy, part of the International Union of Geological Sciences.

19. William Smith (1769-1839). For more on Smith and his maps see *STRATA: William Smith's Geological Maps*. London: Thames & Hudson, 2020; an excellent and extensive online resource is www.strata-smith.com. The best summary of the development of geology is Rudwick, M.J.S., 2014. *Earth's Deep History. How it was discovered and why it matters*. Chicago and London: University of Chicago Press; for more detail see Rudwick, M.J.S., 2005. *Bursting the limits of time. The Reconstruction of Geohistory in the Age of Revolution*. Chicago and London: University of Chicago Press; and Rudwick, M.J.S., 2008. *Worlds before Adam. The Reconstruction of Geohistory in the Age of Reform*. Chicago and London: University of Chicago Press.

20. William Buckland (1784-1856); William Daniel Conybeare (1787-1857); Henry Thomas De la Beche (1796-1855). Hugh Torrens has reviewed the history of geology in the county of Dorset in Torrens, H.S., 2004. How the long history of geological studies in Dorset confirms its World Heritage Coast status. *Open University Geological Society Journal*, v.25, 1-16.

21. The Geological Society of London was founded in 1807, the Bristol Philosophical and Literary Society founded in 1820 opened its Institution for the Advancement of Science, Literature and the Arts in 1823, and the Bath Literary and Scientific Institution also opened in 1823. For more on science in the West Country, see Eyles, V.A., 1955. Scientific activity in the Bristol region in the past. In MacInnes, C.M. & Whittard, W.F. (eds.) *Bristol and its Adjoining Counties*. Bristol: British Association for the Advancement of Science, 123-143. For more on the Geological Society, see Lewis, C.L.E. & Knell, S.J. (eds) *The Making of the Geological Society of London*. The Geological Society, London, Special Publications, No.317.

2. THE GIRL WHO LIVED

1. For more on food riots see Charlesworth, A. (ed.), 2018. *An atlas of rural protest in Britain 1548-1900*. London and New York: Routledge. For a local perspective on social change and rural unrest in Dorset in the early nineteenth century see Draper, J., 2000. *Regency, Riot and Reform*. Wimborne: The Dovecote Press.

2. From 1795 to 1797, the Wordsworths stayed at Racedown Lodge, a house belonging to the Pinney family.

3. Molly Moores was the daughter of Robert Moors and Mary Mitchell, of Durweston near Blandford. Molly and Richard settled in Lyme, although they seem to have moved back and forth, initially at least, and perhaps seasonally, between there and Richard's home village of Sidbury. Richard was not the only Anning to settle in Lyme; others, probably his cousins, came from Colyton in Devon to Lyme, one of whom, Simeon Anning worked as a baker and confectioner. For this information on the Anning family history I am much indebted to Professor Hugh Torrens whose relentless pursuit of the Annings over many years has located their origins and added much interesting detail.

4. This second Henry was baptised on 28 March 1802 but was buried just three days later; Perceval was baptised on 30 March 1804 and buried on 5 April 1804; and Elisabeth, born on 24 December 1805 was buried on 31 March 1806.

5. Reverend John Crook (*c.*1766-1830) was officiating because Lyme's minister at the time, the Reverend John Reed Harris, had been removed from his post in 1798 and moved to Ilminster where he became minister of the Socinian chapel.

6. Reverend James Wheaton (1773-1818). For more on the history of the Independent chapel in Coombe Street and James Wheaton, see Densham, W. & Ogle, J., 1899. *The story of the Congregational churches of Dorset: from their foundation to the present time*. Bournemouth: Mate & Sons. For details of the building see *An inventory of the historical monuments in Dorset, volume 1, West*. London, HMSO, 1952.

7. For a consideration of Mary Anning's religious faith see Goodhue, T.W., 2001. The faith of a fossilist: Mary Anning. *Anglican and Episcopal History*, v.70, 80-100; Goodhue, T.W., 2005. Mary Anning: the fossilist as exegete. *Endeavour*, v.29, 28-32; and Goodhue, T.W., 2004. *Fossil hunter. The life and times of Mary Anning (1799-1847)*. Bethesda, Maryland: Academic Press. The volume of *The Theological Magazine* is signed 'Mary Anning Her Book, January 15, 1807' (possibly 1809).

8. O'Connor, R., 2009. Facts and fancies: the Geological Society of London and the wider public, 1807-1837. *In* Lewis, C.L.E. & Knell, S.J. (eds), *The Making of the Geological Society of London*. The Geological Society, London, Special Publications, No. 317, 334, note 6.

9. Reports of the lightning strike are in various newspapers such as the *Kentish Mercury*, Friday 29 August 1800, 2; *Northampton Mercury*, Saturday 30 August 1800, 3; *Newcastle Courant*, Saturday 30 August 1800, 3-4; *Ipswich Journal*, Saturday 30 August 1800, 4; and *Sussex Advertiser*, Monday 1 September 1800, 1. John Hasking's 1821 account of Mary Anning being struck by lightning was sent by Charles Churchill Anning, Mary's nephew, to Henry De la Beche in July 1847 and is preserved in the De la Beche Archive in the National Museum of Wales (NMW84.20G.D11). See Lang, W.D., 1958. Mary Anning's escape from lightning. *Proceedings of the Dorset Natural History & Archaeological Society*, v.80, 91-93.

10. Draper, J., 2004. *Mary Anning's town – Lyme Regis*. Lyme Regis: Lyme Regis Museum, 48. I am grateful to Richard Bull of Lyme Regis Museum for more details of events at the Rack Field, including the reference to Mr Sand's equestrian display which is from the *Dorset County Chronicle*, 15 June 1843.

11. I am most grateful to Mike Taylor for kindly supplying information about the properties occupied by the Annings. Cockmoile is also occasionally spelt 'Cockmoil' on maps of the period.

12. James Johnson (c.1764-1844); William Cunnington (1754-1810). James Johnson's letter of 23 July 1810 is in the William Cunnington MSS, Wiltshire Archaeological Society Library in Devizes Museum. 'Mr Carpenter' is Dr Thomas Coulson Carpenter, (c.1775-1833). According to the inquest report on the death of Elizabeth Haskings, it was Carpenter who attended her and pronounced her dead. It may have been he who suggested the method of reviving the infant Mary.

13. Roberts, G., 1834. *The History and Antiquities of the Borough of Lyme Regis and Charmouth*. London: Samuel Bagster, 286.

14. Reverend Stebbing Shaw (1762-1802). Shaw's meeting with William Lloyd (or Lock?) in 1788 is on page 451 of Shaw, S., 1789. *A tour to the West of England in 1788*. London. W.D. Lang in his 1960 paper, Portraits of Mary Anning and other items. *Proceedings of the Dorset Natural History & Archaeological Society*, v.81, 89-91 records this meeting as taking place in 1739 from handwritten notes

by George Roberts, probably a mis-reading of the date.

15. George Roberts' description of the early fossil collectors is on page 557 of Roberts, G., 1856. *The social history of the people of the southern counties of England in past centuries; illustrated in regard to their habits, municipal bye-laws, civil progress, etc.* London: Longman, Brown, Green, Longmans, & Roberts; and on pages 286-287 of Roberts, G., 1834. *The History and Antiquities of the Borough of Lyme Regis and Charmouth.* London: Samuel Bagster.

16. William Lock (c.1739-1814). William George Maton (1774-1835). Maton's meeting with William Lock at Charmouth is on pages 75-79 of volume 1 of Maton, W.G., 1797. *Observations relative chiefly to the natural history, picturesque scenery and antiquities of the western counties of England, made in the years 1794 and 1796.* Salisbury: J. Easton: 'All curious productions of this nature are diligently collected by a man living at Charmouth who is generally known throughout the county by the name of Curi-man. We purchased from his collection some fine specimens of chalcedony'. Seemingly, we have two local men in Charmouth with similar names recorded as collecting and selling curios to travellers, William Lock and William Lloyd. Are they the same person? Did Stebbing Shaw mishear the name as 'Loyd' when he met William Lock?

17. John South (c.1778-1831). South was collecting actively until at least 1813.

18. John Crookshanks (c.1738-1802). Roberts, G., 1834. *The History and Antiquities of the Borough of Lyme Regis and Charmouth.* London: Samuel Bagster, 286.

19. From the limited records we have, it can be difficult to determine whether a particular collector was an active searcher of fossils or a purchaser, which becomes significant when trying to judge the contribution of that individual. How much did, for example, James Johnson collect himself, and how much did he purchase?

20. Roberts, G., 1823. *The History of Lyme Regis, Dorset, from the Earliest Periods to the Present Day.* Sherborne: Langdon & Harker, 122.

21. Mary Anning to Charles Lyell, 15 December 1829. Copy in Lyme Regis Museum.

22. Jean André De Luc (1727-1817). De Luc's account of his visit to Lyme is in De Luc, J.A., 1811. *Geological travels, volume II: Travels in England. Translated from the French manuscript.* London: F.C. & J. Rivington. See also Lang, W.D., 1939. Mary Anning (1799-1847) and the pioneer geologists of Lyme. *Proceedings of the Dorset Natural History & Archaeological Society*, v.60, 142-164.

23. Threat to the churchyard by erosion was still present in 1840, when Lyme resident Sarah Benett, wife of Captain Charles Cowper Benett RN (1789-1878) wrote to Henry De la Beche, then Director of the Geological Survey, asking him to 'inspect the damage before we take any steps to remedy it', Sarah Benett to H.T. De la Beche, 25 April 1840, De la Beche Archive, National Museum of

Wales, NMW84.20G.D80. However, nothing seems to have been done until the early twentieth century, by which time human bones and skulls were eroding out of the churchyard graves. Work to protect Church Cliffs was undertaken in 1911, but extensive repairs were needed in 1948 and again in 1953. Construction of a new, extended sea wall and stabilisation of the cliffs took place in 2012-14 at, as De Luc pointed out in 1811, 'a great deal of expense'. For more on the recent stabilisation of Church Cliffs see Gallois, R.W., 2016. Geological investigations for coastal protection and landslide remedial works at Lyme Regis, Dorset, UK. *Geoscience in South-West England*, v.14, 1-11.

24. The description of Richard Anning's fall is in Roberts, G., 1834. *The History and Antiquities of the Borough of Lyme Regis and Charmouth*. London: Samuel Bagster, 288. Richard junior was the Anning's fourth child, born about 1797, but who died aged 14, less than a year after his father, and was buried on 12 August 1811. Confusingly, there is a record of another Richard, perhaps the Anning's tenth child, born on 30 September 1810, baptised a week after his father's burial, and himself buried a year later, on 8 December 1811. If Molly did indeed bear another son in September 1810, she would have been about 47 years old when he was born.

3. THE FIRST ICHTHYOSAURS

1. Roberts, G., 1834. *The History and Antiquities of the Borough of Lyme Regis and Charmouth*. London: Samuel Bagster, 288.

2. Anna Maria Pinney (1812-1861). Lang, W.D., 1956. Mary Anning and Anna Maria Pinney. *Proceedings of the Dorset Natural History & Archaeological Society*, v.76, 146. One wonders who the 'fossilist' purchaser of this first specimen was; might it have been one of three sisters, the Misses Philpot, who had moved to Lyme about five years earlier and who seem to have taken to fossil collecting around the time of Richard Anning's death?

3. Grant, J., 1827. *A memoir of Miss Francis August Bell, who died in Kentish Town, on Monday the 23rd of May, 1825, aged 15 years and six months: with specimens of her compositions, in prose and verse*. London: Hatchard & Son.

4. Charles Churchill Anning to H.T. De la Beche, July 1847. De la Beche Archive, National Museum of Wales, NMW84.20G.D11.

5. *The Western Flying Post; or Sherborne and Yeovil Mercury*, Monday November 9, 1812; the discovery was also reported in the *Manchester Mercury*. Tuesday 17 November 1812, 4; and *Royal Cornwall Gazette*, Saturday 19 December 1812, 4.

6. *Sussex Advertiser*, Monday 20 December 1813, 3; *Northern Mercury*, Saturday 25 December 1813, 3; *Caledonian Mercury*, Monday 27 December 1813, 1; *Liverpool Mercury*, Friday 31 December 1813, 7. *The Monthly*

Magazine, or British Register, v.37, 1 February 1814, 94.

7. In 1847 Charles Churchill Anning was thirteen years old and writing about events which had happened thirty-five years earlier and over twenty years before he was born. The information, however, must have come from his father who had the opportunity, perhaps for the first time, to correct the long-held view that it was his sister who discovered the famous specimen.

8. See Young, G., 1825. Account of a Fossil Crocodile recently discovered in the Alum-Shale near Whitby, *The Edinburgh Philosophical Journal,* v.13, 76-81 for a discovery of a fossil crocodile in Whitby in December 1824. Rev. Peter Hawker (c.1773-1833). Hawker, J., *Gentleman's Magazine,* v.77, 7-8. For more on the history of the discovery of ichthyosaurs, see Delair, J.B., 1969. A history of the early discoveries of Liassic ichthyosaurs in Dorset and Somerset (1779-1835). *Proceedings of the Dorset Natural History and Archaeological Society,* v.90, 115-127; Howe, S.R., Sharpe, T. & Torrens, H.S., 1981. *Ichthyosaurs: a history of fossil 'sea-dragons'.* Cardiff: National Museum of Wales; and Evans, M., 2010. The roles played by museums, collections and collectors in the early history of reptile palaeontology. *In* Moody, R.T.J., Buffetaut, E., Naish, D. & Martill, D.M. (eds), *Dinosaurs and Other Extinct Saurians: a Historical Perspective.* The Geological Society, London, Special Publications. No.343, 5-29.

9. Henry Hoste Henley (1766-1833). The comparison to the wages of Richard Anning is from Noè, L., Gómez-Pérez, M. & Nicholls, R., 2019. Mary Anning, Alfred Nicholson Leeds and Steve Etches. Comparing the three most important UK 'amateur' fossil collectors and their collections. *Proceedings of the Geologists' Association,* v.130, 376.

10. Charlotte Stock (c.1779-1861); Dr John Edmonds Stock (c.1774-1835). More about Mrs Stock can be found in Torrens, H.S., 2017. Appendix 1. The mystery of Dr John Edmonds Stock, Beddoes' first biographer. *In* Levere, T., Stewart, L., Torrens, H.S. & Wachelder, J., *The Enlightenment and Thomas Beddoes, Science, medicine, reform.* Science, Technology and Culture 1700-1945. Abingdon: Routledge, 238-248. Robert Bakewell (1767-1843). His *Introduction to Geology, illustrative of the General Structure of the Earth; comprising the Elements of the Science, and an Outline of the Geology and Mineral Geography of England* was one of the first geological textbooks. First published in 1813, it went through five editions by 1838 as well as a German and several American editions.

11. Lieutenant-Colonel Thomas James Birch (later Bosvile) (1768-1829).

12. William Bullock (1773-1849). Charles Konig (1774-1851), Keeper of Natural History at the British Museum. The Anning skull is listed on p.96 of Bullock, W., 1816. *A companion to the London Museum, and Pantherion, containing a brief description of upwards of fifteen thousand natural and foreign curiosities, antiquities, and productions of the fine arts; now open for public*

inspection in The Egyptian Temple, Piccadilly, London. 17th edition. London: Printed for the Proprietor. The auction catalogue of Bullock's sale with manuscript prices and buyers' names was reprinted in facsimile in 1979: *Sale catalogue of the Bullock Museum 1819.* London: Harmer Johnson & John Hewitt.

13. Sir Everard Home (1756-1832). Home, E., 1814. Some account of the fossil remains of an animal more nearly allied to fishes than to any of the other classes of animals. *Philosophical Transactions of the Royal Society of London,* v.104, 572.

14. William Buckland to Charles Konig 3 April 1835, Natural History Museum Library. Roberts, G., 1834. *The History and Antiquities of the Borough of Lyme Regis and Charmouth.* London: Samuel Bagster, 288. Anning's association with this first ichthyosaur discovery was remembered also by the *Gloucester Journal,* Saturday 20 May 1843, but refers to her as 'Mary Ann Anning. Armed with her geological hammer, she battled it with beds of lias till she dislodged from its quiet resting-place between Lyme and Charmouth, in the year 1814, the first fossil remains of an Ichthyosaurus.'

15. Home published five further papers in the *Philosophical Transactions* between 1816 and 1820: Home, E., 1816. Some farther account of the fossil remains of an animal, of which a description was given to the Society in 1814. *Philosophical Transactions of the Royal Society of London,* v.106, 318-321; Home, E., 1818. Additional facts respecting the fossil remains of an animal, on the subject of which two papers have been printed in the Philosophical Transactions, showing that the bones of the sternum resemble those of the Ornithorhynchus Paradoxus. *Philosophical Transactions of the Royal Society of London,* v.108, 24-32; Home, E., 1819. An account of the fossil skeleton of the Proteo-Saurus. *Philosophical Transactions of the Royal Society of London,* v.109, 209-211; Home, E., 1819. Reasons for giving the name Proteo-Saurus to the fossil skeleton which has been described. *Philosophical Transactions of the Royal Society of London,* v.109, 212-216; and Home, E., 1820. On the mode of formation of the canal for containing the spinal marrow, and on the form of the fins (if they deserve that name) of the Proteosaurus. *Philosophical Transactions of the Royal Society of London,* v.110, 159-164.

16. The account James Johnson's paper to the Linnean Society is to be found in *Annals of Philosophy,* 1815, v.5, 70 and in *The Scots Magazine and Edinburgh Literary Miscellany,* February 1815, v.77, 133 and in other newspapers of the time such as the *Hereford Journal.*

17. Roberts, G., 1856. *The Social History of the People of the Southern Counties of England in Past Centuries.* London: Longman, Brown, Green, Longmans, & Roberts, 557.

18. Jean-Léopold-Nicholas-Frédéric Cuvier (1769-1832), Baron Cuvier from 1819; he adopted the name Georges after the death of his elder brother Georges

Charles Henri.

19. Charles Lyell (1797-1875). *Principles of geology, being an attempt to explain the former changes of the Earth's surface, by reference to causes now in operation* was published in three volumes between 1830 and 1833 by John Murray. James Hutton (1726-1797). The quote is the final line of Hutton, J., 1788. Theory of the Earth. *Transactions of the Royal Society of Edinburgh*, v.1, 304. The idea of slow, gradual change resulting from processes acting on the earth is known as uniformitarianism.

20. For more on Cuvier see Rudwick, M.J.S., 1997. *Georges Cuvier, Fossil Bones, and Geological Catastrophes. New Translations & Interpretations of the Primary Texts*. Chicago and London: University of Chicago Press. For more on the discovery of fossil marine reptiles and their terrestrial counterparts, the dinosaurs, see McGowan, C., 2001. *The Dragon Seekers*. Cambridge MA: Perseus; Cadbury, D., 2000. *The Dinosaur Hunters*. London: Fourth Estate; Knell, S.J., 2000. *The culture of English Geology, 1815-1851. A science revealed through its collecting*. Aldershot: Ashgate; and Freeman, M., 2004. *Victorians and the Prehistoric. Tracks to a Lost World*. New Haven & London: Yale University Press.

4. FRIENDS AND NEIGHBOURS

1. (Later Sir) Henry Thomas De la Beche (1796-1855). Elizabeth De la Beche née Smith (c.1779-1833); William Huddle Aveline (1771-1837). For more on Henry De la Beche see McCartney, P.J., 1977. *Henry De la Beche: observations on an observer*. Cardiff: Friends of the National Museum of Wales; Sharpe, T., 2013. New insights into the early life of Henry Thomas De la Beche (1796-1855). *In* Morris, R. (ed.) *A journal of Sir Henry De la Beche pioneer geologist (1796-1855) written in his own hand*. Swansea: Royal Institution of South Wales, 5-21; and Bate, D.G., 2010. Sir Henry Thomas De la Beche and the founding of the British Geological Survey. *Mercian Geologist*, v.17, 149-165. I am grateful to Lyme historian Keith Shaw for information on Aveline's occupancy of Woodville.

2. It is from Mary Anning that we have an indication of the pronunciation of De la Beche's name. Her spelling is often phonetic, and when she mentions him in correspondence, it is almost always as 'Mr De la Beach', which we can safely assume is how she said his name, and how he himself pronounced it. The family name was, in fact, Beach until it was changed by Henry's father and uncle in 1790, but Mary Anning is unlikely to have known that. From his descendants, although none bears that name now, we learn that it should be pronounced as 'Della Beach'.

3. George Holland (1787-1858).

4. Etheldred Benett (1776-1845). The quote from Benett about De la Beche is in a letter of 1818 to George Bellas Greenough in Cambridge University Library.

De la Beche's income was derived from a slave-worked sugar estate in Jamaica which his father had inherited. When it passed, heavily mortgaged, to De la Beche, who detested slavery, he found himself in a difficult position, but through conditions in his father's will he was unable to dispose of it. He did what he could to improve conditions on the estate in the face of opposition from other estate owners on the island. He supported the idea of a gradual abolition of slavery.

5. The name '*Ichthyosaurus*' first appears in Konig, C., 1817. *Synopsis of the contents of the British Museum.* 11th edition. London.

6. Charles Hunnings Wilkinson (c.1763-1850). De la Beche's journals for this period are in the National Museum of Wales. See Sharpe, T. & McCartney, P.J., 1998. *The papers of H.T. De la Beche (1796-1855) in the National Museum of Wales.* Cardiff: National Museum of Wales.

7. De la Beche, H.T., 1824. Remarks on the geology of the south coast of England, from Bridport Harbour, Dorset to Babbacombe Bay, Devon. Read 5 March 1819. *Transactions of the Geological Society of London, Second Series,* v.1, 43. *Ichthyosaurus platyodon* was later renamed *Temnodontosaurus platyodon* by palaeontologists Henry Alleyne Nicholson (1844-1899) and Richard Lydekker (1849-1915) in their 1889 book, *A Manual of Palaeontology for the Use of Students with a General Introduction on the Principles of Palaeontology.* Edinburgh and London: William Blackwood, third edition, v.2, xi for the shape of its teeth, the name meaning 'cutting-tooth lizard'. *Ichthyosaurus tenuirostris* which had been renamed *Leptopterygius* in 1922 was put into a new genus, *Leptonectes*, by Chris McGowan, 1996. The taxonomic status of *Leptopterygius* Huene, 1922 (Reptilia: Ichthyosauria). *Canadian Journal of Earth Sciences,* v.33, 439-443.

8. William Buckland (1784-1856); William Daniel Conybeare (1787-1857). De le Beche's meeting with Conybeare is from Torrens, H.S., 2004. Conybeare, William Daniel (1787-1857). *Oxford Dictionary of National Biography;* and his meeting with Buckland from Buckland, F.T., 1858. Memoir of the Very Rev. William Buckland, D.D., F.R.S., Dean of Westminster. *In* Buckland, W., *Geology and Mineralogy as exhibiting the Power, Wisdom, and Goodness of God.* Third edition, xxxi.

9. For more on Buckland's Axminster connections see Powell, C., 2018. *Lyme Regis Monographs.* FeedARead.com Publishing, 278-294. The quote by his son is from Buckland, F.T., 1858, Memoir of the Very Rev. William Buckland, D.D., F.R.S., Dean of Westminster. *In* Buckland, W., *Geology and Mineralogy as exhibiting the Power, Wisdom, and Goodness of God.* Third edition, xxvii. The quote about Buckland's visits to Lyme is from Gordon, Mrs [Elizabeth Oke Buckland], 1894. *The life and correspondence of William Buckland, D.D., F.R.S.* London: John Murray, 3, 7-8. In these recollections, written towards the end of the nineteenth century, Elizabeth misremembers Mary Anning's name and places

NOTES, SOURCES AND FURTHER READING

Buckland's outings with Mary earlier than they were. Rather than being during his early years at Oxford, these excursions were more likely to have taken place from the 1820s onwards.

10. John Kidd (1775-1851). For more on Buckland see Haile, N., 2004. Buckland, William (1784-1856. *Oxford Dictionary of National Biography;* Rupke, N.A., 1983. *The great chain of history. William Buckland and the English School of Geology (1814-1849).* Oxford: Clarendon Press. Darwin's opinion of Buckland is in Barlow, N. (ed). 1958. *The autobiography of Charles Darwin 1809-1882. With the original omissions restored. Edited and with appendix and notes by his grand-daughter Nora Barlow.* London: Collins, 102. Buckland's admission that he felt nervous is from a remark made in 1836 and recorded by Caroline Fox (1819-1871) in her journal; see Pym, H. (ed.), 1882. *Memories of Old Friends being extracts from the Journals and Letters of Caroline Fox of Penjerrick, Cornwall from 1835 to 1871.* London: Smith, Elder, & Co, 5.

11. Charles Lyell to Gideon Mantell, 8 February 1822, quoted in Lyell, Mrs [K.M.] (ed.), 1881. *Life letters and journals of Sir Charles Lyell, Bart.* 2 vols. London: John Murray, v.1, 115.

12. Buckland had his inaugural lecture published in 1820 with the title *Vindiciae Geologicae; or the Connexion of Geology with religion explained.* For more on this see Edmonds, J.M., 1991. *Vindiciae Geologicae,* published 1820; the inaugural lecture of William Buckland. *Archives of Natural History,* v.18, 255-268. For more on clergy-scientists and natural theology, see Chapman, A., 2020. *Caves, Coprolites, and Catastrophes. The Story of Pioneering Geologist and Fossil-Hunter William Buckland.* London: Society for Promoting Christian Knowledge.

13. Charles Lyell to Gideon Mantell, 20 July 1825, quoted in Lyell, Mrs [K.M.] (ed.), 1881. *Life letters and journals of Sir Charles Lyell, Bart.* 2 vols. London: John Murray, v.1, 161.

14. This description of Mrs Buckland is from Pym, H. (ed.), 1882. *Memories of Old Friends being extracts from the Journals and Letters of Caroline Fox of Penjerrick, Cornwall from 1835 to 1871.* London: Smith, Elder, & Co, 44.

15. Richard Whately (1787-1863), a contemporary of Buckland's at Oxford. For a full version of Whately's *Elegy* see Gordon, Mrs [Elizabeth Oke Buckland], 1894. *The life and correspondence of William Buckland, D.D., F.R.S.* London: John Murray, 41-42 Buckland was delighted with Whately's poem and had copies printed which he distributed to friends.

16. Willy's elder brother and mentor was John Josias Conybeare (1779-1824) whom he joined at Christ's Church. Following the death of his brother in 1824, although he maintained an interest, Willy's active involvement in geological matters waned and he became more focussed on his church responsibilities. For more on William Daniel Conybeare see North,. F.J., 1933. Dean Conybeare,

geologist. *Transactions of the Cardiff Naturalists' Society*, v.66, 15-68; North, F.J., 1956. W.D. Conybeare, his geological contemporaries and Bristol Associations. *Proceedings of the Bristol Naturalists' Society*, v.29, 133-146; Powell, C., 2018. *Lyme Regis Monographs*. FeedARead.com Publishing, 226-277; Torrens, H.S., 2004. Conybeare, William Daniel (1787-1857). *Oxford Dictionary of National Biography*.

17. William Phillips (1775-1828); Conybeare, W.D. and Phillips, W., 1822. *Outlines of the Geology of England and Wales, Part One*. London: William Phillips. Only this first part was ever published. This book was an update of Phillips' 1818 *A selection of Facts from the Best Authorities, arranged so as to form an Outline of the Geology of England and Wales*.

18. The three Philpot sisters were Mary (1777-1838), Elizabeth (1780-1857) and Margaret (1786-1845); their brother was John Philpot (1778-1850). Letters from William and Mary Buckland to Elizabeth Philpot, 1 January 1834, private collection.

19. The description of the Philpot collection is from Salina Hallett (*born c.*1839), 1993. *Lyme voices I*. Lyme Regis: Lyme Regis Museum.

20. A notice of Elizabeth Philpot's death in 1857 records that 'Miss Philpot went upon the lias shore in company with Mary Anning almost daily'. *The Athenaeum*, no.1556, 22 August 1857, 1059.

21. Sowerby, J. & Sowerby J. De C., 1812-46. *The Mineral Conchology of Great Britain*. Seven volumes. London. A list of fossils from Charmouth described by the Sowerbys and the account of Elizabeth Philpot trying to have her specimen returned to her is in Lang, W.D., 1941. Early days of natural history at Charmouth. *Proceedings of the Dorset Natural History & Archaeological Society*, v.62, 113. The Charmouth fossils described include *Ammonites henleyi*, named after the local lord of the manor, H.H. Henley; *Ammonites bechei*, after De la Beche; and *Ammonites birchi* after Colonel Thomas Birch.

22. Ann Congreve (1758-1823) and Mary Congreve (1756-1823). Lyme burial records note that Mary was buried on 4 April 1823 and Ann on 2 January 1824. Ann was a subscriber and donor to several Christian organisations. Ann's will records her wishes for her fossil collection but its destiny is so far unknown.

23. Sarah Kennaway, née Johnson (1767-1855). Her husband Robert (*c.*1761-1829) was from a long-established family of Exeter wool and, later, wine merchants and was in partnership with Thomas Finimore Hill (1783-1869) until that was dissolved in 1811. Robert was a subscriber to George Roberts' 1823 *History of Lyme Regis*. Sarah was also a hobby poet; see Press, P.M. & Priestley, M., 2012. Sarah Kennaway. *The Village Echo. The Journal of the Pavey Group*, no.38, 19-25. Robert's death in 1829 left Sarah a 'lady of means' who contributed to various charitable causes and remained in Charmouth until her death in 1855 when she was buried alongside her husband in the churchyard of St Andrew's in

NOTES, SOURCES AND FURTHER READING

Charmouth. The letter from Mary Anning to Sarah Kennaway is in the Archives of the British Geological Survey.

24. For more on Birch see: Torrens, H.S., 1979. Collections and collectors of note. 28. Colonel Birch (c.1768-1829). *Newsletter of the Geological Curators' Group*, v.2, 405-412; Torrens, H.S., 1980. Collections and collectors of note. 28. Colonel Birch (c.1768-1829). *The Geological Curator*, v.2, 561-562. Home's letter of 23 September 1818 to De la Beche is in the De la Beche Archive at the National Museum of Wales, NMW84.20G.D699.

5. THE FIRST PLESIOSAURS

1. Dr Gideon Algernon (1790-1852). Birch's visit to Mantell is noted in Cooper, J.A. (ed.), 2010. *The unpublished journal of Gideon Mantell 1819-1852.* Brighton & Hove: The Royal Pavilion & Museums. The 5 March 1820 letter from Birch to Mantell is in the Mantell MSS, Alexander Turnbull Library, Wellington, New Zealand. Mantell's comment that the best fossils had been obtained by Mary Anning is from Mantell, G.A, 1846. A few notes on the prices of fossils. *London Geological Journal*, v.1, 13.

2. Details of some of the prices reached are published by Mantell, G.A, 1846. A few notes on the prices of fossils. *London Geological Journal*, v.1, 13-17.

3. George Cumberland (1754-1848). Cumberland's 20 August 1820 implication of some romantic entanglement of Birch and Mary Anning is in British Library Add MSS 36520.

4. For more on Sedgwick see Speakman, C., 1982. *Adam Sedgwick Geologist and Dalesman 1785-1873. A biography in twelve themes.* Heathfield: The Broad Oak Press. Sedgwick's knowledge of geology was perhaps not as basic as is often suggested. Although he did not join the Geological Society until 1818, he had attended meetings as a visitor for at least two years prior.

5. The Cambrian Period extends from 541 to 485 million years ago; the Devonian from 419 to 359 million years ago.

6. For more on Sedgwick's purchases from Anning see Price, D., 1986. Mary Anning specimens in the Sedgwick Museum, Cambridge. *The Geological Curator*, v.4, 319-324. For newspaper reports see for example the *Public Ledger and Daily Advertiser*, Saturday 26 May 1821, 1-2: '... These interesting vestiges of the primeval world are now in the possession of Miss Mary Anning, of Lyme, who discovered these extraneous fossils.'

7. The 1822 letter from Anning to Sarah Kennaway is in the Archives of the British Geological Survey.

8. The Lyme Fossil Shop's copy of Birch's sale catalogue is in the Library of the Natural History Museum.

9. George Weare Braikenridge (1775-1856); James Johnson (c.1764-1844);

Richard Bright (1754-1840); William Morgan (c.1773-1852); Charles Hunnings Wilkinson (c.1763-1850); Samuel Skurray Day (1787-1816). The remark that Johnson's collection was deficient was made by George Cumberland: see Knell, S.J., 2000. *The culture of English Geology, 1815-1851. A science revealed through its collecting*. Aldershot: Ashgate, 73-74.

10. More on the early discoveries of marine reptiles can be found in: Howe, S.R., Sharpe, T. & Torrens, H.S., 1981. *Ichthyosaurs: a history of fossil 'sea-dragons'*. Cardiff: National Museum of Wales; and Evans, M., 2010. The roles played by museums, collections and collectors in the early history of reptile palaeontology. *In* Moody, R.T.J., Buffeteaut, E., Naish, D. & Martill, D.M. (eds), *Dinosaurs and other extinct saurians: a historical perspective*. The Geological Society, London, Special Publications, No.343, 287-311. Cumberland, G., 1829. Some account of the order in which the fossil saurians were discovered. *Quarterly Journal of Science, Literature and Art*, v.27 (New Series 5), 345-349. That Everard Home was displeased with the ichthyosaur paper is from Buckland to De la Beche 21 November 1821, National Museum of Wales, De la Beche Archive NMW84.20G. D161. Conybeare's expectation of this response from Home is in a letter to Buckland, March 1821, in the archives of the Royal Society MS/251/20.

11. At least part of Birch's plesiosaur specimen illustrated by De la Beche and Conybeare has survived and is in the collections of Oxford University Museum of Natural History (OUMNH J.50146: see Evans, M., 2010. The roles played by museums, collections and collectors in the early history of reptile palaeontology. *In* Moody, R.T.J., Buffeteaut, E., Naish, D. & Martill, D.M. (eds), *Dinosaurs and other extinct saurians: a historical perspective*. The Geological Society, London, Special Publications, No.343, 16.

12. De la Beche, H.T. & Conybeare, W.D., 1821. Notice of the discovery of a new fossil animal, forming a link between the Ichthyosaurus and the crocodile; together with general remarks on the osteology of the Ichthyosaurus. Read 6 April 1821. *Transactions of the Geological Society of London, First Series*, v.5, 559-594. Alexander Catcott (1725-1779); William Stukeley (1687-1765).

13. Conybeare's amended manuscript of their paper is in the De la Beche Archive in the Department of Geology, National Museum of Wales, NMW84.20G.D300. Conybeare must have received some criticism over the name *Plesiosaurus*, and in a later paper was obliged to justify his invention of it. He was on firm ground: he did, after all, have a First Class degree in classics from Oxford University. His justification of the composition of the name *Plesiosaurus* is explained in a footnote on page 381 of his 1824 paper On the Discovery of an almost perfect Skeleton of the Plesiosaurus. *Transactions of the Geological Society of London, Second Series*, v.1.

14. Thomas Clarke (1792-1864). The plesiosaur skull found by Clarke was presented to the Geological Society in 1823. It is now in the collections of

NOTES, SOURCES AND FURTHER READING

the British Geological Survey (BGS GSM 26035) and has been redescribed as *Thalassiodracon hawkinsi*. Conybeare, W.D., 1822. Additional notices on the fossil genera Ichthyosaurus and Plesiosaurus. Read May 3, 1822. *Transactions of the Geological Society of London, Second Series*, v.1, 103-123.

15. The reports of Mary Anning's 1821 discoveries of ichthyosaurs are in *The Monthly Magazine or British Register*, 1821, v.51, 553; *The New Monthly Magazine and Literary Journal*, 1821, v.3, 339; and *The New Monthly Magazine and Literary Journal*, 1829, part iii. Historical Register, 43.

16. *The New Monthly Magazine and Literary Journal*, 1822, v.6, Historical Register., 420.

17. *The Monthly Magazine; or British Register*, vol LX, Part II for 1825. Oct 1, 286.

18. *The New Monthly Magazine and Literary Journal*, 1828, July 1, 323.

19. De la Beche's letter to Konig of 11 July 1821 is quoted in Torrens, H.S., 1995. Mary Anning (1799-1847) of Lyme; 'the greatest fossilist the world ever knew'. *British Journal for the History of Science*, v.28, 262.

20. Molly Anning's letter to Konig of 2 September 1821 is in the Natural History Museum Archives DF100/1/2. The specimen discussed in this letter is probably that catalogued as NHMUK R808 and 809. See Chapman, S. & Milner, A.C., 2010. Appendix. A catalogue of fossil specimens collected by Mary Anning (1799-1847) held in the collections at the Natural History Museum, London. *In* Lord, A.R. & Davis, P.G. (eds), 2010. *Fossils from the Lower Lias of the Dorset coast*. Palaeontological Association Field Guide to Fossils: Number 13. London: The Palaeontological Association, 403.

21. This ichthyosaur, purchased for the Bristol Institution, was destroyed during a bombing raid on Bristol in November 1940.

22. Joseph Anning's letter of 13 September 1825 is in the Archives of the Natural History Museum, DF100/1/1, and the specimen described is now in the Museum's Fossil Marine Reptile Gallery, registration number NHMUK R 1120. See Chapman, S. & Milner, A.C., 2010. Appendix. A catalogue of fossil specimens collected by Mary Anning (1799-1847) held in the collections at the Natural History Museum, London. *In* Lord, A.R. & Davis, P.G. (eds), 2010. *Fossils from the Lower Lias of the Dorset coast*. Palaeontological Association Field Guide to Fossils: Number 13. London: The Palaeontological Association, 402.

23. George Cumberland's description of the hazards of Mary Anning's work is in the *Bristol Mirror*, 11 January 1823, 4 and quoted in Torrens, H.S., 1995. Mary Anning (1799-1847) of Lyme; 'the greatest fossilist the world ever knew'. *British Journal for the History of Science*, v.28, 257-284.

24. Roberts, G., 1823. *The History of Lyme Regis, Dorset, from the Earliest Periods to the Present Day*. Sherborne: Langdon & Harker. Anning's copy was

located in Cardiff University Library by Chris Powell in 2005 and I am grateful to him for this information. For an idea of the prices of some books of this period and their equivalents in relation to the weekly wage of a lawyer's clerk, see O'Connor, R., 2007. *The Earth on show. Fossils and the poetics of popular science, 1802-1856*. Chicago and London: University of Chicago Press, 219-223.

25. John Murray (c.1786-1851); See Murray, J., 1847. The late Miss Mary Anning. *Mining Journal*, v.17, 25 December 1847.

26. Lang, W.D., 1956. Mary Anning and Anna Maria Pinney. *Proceedings of the Dorset Natural History & Archaeological Society*, v.76, 146-152.

27. Roberts, G., 1839. *An etymological and explanatory dictionary of the terms and language of geology designed for the early student, and those who have not made great progress in that science*. London: Longman, Orme, Brown, Green, & Longmans, 84.

28. According to some 1892 reminiscences, by 'X', *More Memories of Old Lyme*. Lyme Regis: J.S. Turner, 'A large quantity of cement stone was conveyed on the backs of donkeys, and every morning a drove of from twenty to thirty of these poor illused animals, fresh from their pasture and nights repose, scampered down Broad Street, rearing, kicking, and prancing followed by a crowd of boys, hooting and yelling to the great diversion of juveniles but the terror of staid ladies.'

29. From the journal of Anna Maria Pinney quoted in Lang, W.D., 1956. Mary Anning and Anna Maria Pinney. *Proceedings of the Dorset Natural History & Archaeological Society*, v.76, 147.

30. Roberts, G., 1834. *The History and Antiquities of the Borough of Lyme Regis and Charmouth*. London: Samuel Bagster, 290.

31. The contemporary report of the discovery is from the *Western Flying Post*, 15 December 1823 and from *The New Monthly Magazine and Literary Journal*, v.12, 92-93, 1 February 1824.

32. Anning notified Sir Henry Bunbury (1778-1860) of her new discovery in a letter of 19 December 1823 and sent him a drawing of it, with her asking price of £110 in a letter of 26 December 1823. MS 8592/1-2, Wellcome Collection, London. The Duke of Buckingham here is Richard Temple-Nugent-Brydges-Chandos-Grenville, (1776-1839), 1st Duke of Buckingham and Chandos.

33. Conybeare's letter of 4 March 1824 to De la Beche is in the De la Beche Archive in the Department of Geology, National Museum of Wales NMW 84.20G.D302.

34. Buckland probably called on Conybeare at his home in Brislington on or about Tuesday 27 January 1824. Conybeare received the drawing from Anning three days later, on the day of a meeting of the Bristol Philosophical Society, probably Friday 30 January. *The Bristol Mirror*, Saturday 31 January 1824, 3 reported the meeting and Conybeare exhibiting 'an illustrative drawing'. It also

notes that on Buckland's recommendation the specimen had been purchased by the Duke of Buckingham. The *Salisbury and Winchester Journal*, Monday 9 February 1824, 2 also carried a report as did the *Dorset County Chronicle*, Thursday 5 February 1824, 4 which noted that 'this hitherto unknown animal was lately discovered at Lyme, by Mary Anning. The plesiosaur, the type specimen of *Plesiosaurus dolichodeirus*, bought by Buckingham was purchased by the British Museum from his estate in 1848 and is now in the Natural History Museum, registration number NHMUK 22656.

35. John Mathew Gutch (1776-1861) was the owner and printer of a local weekly newspaper, *Felix Farley's Bristol Journal*. A report appeared in the 6 March 1824 issue describing 'Mary Anning's New Discovery' as an 'extraordinary fossil fish [which] belongs to Ichthyosaurus tribe', so Conybeare was right to be concerned that there would be some 'strange blunders' as he put it in his letter to De la Beche. See Taylor, M.A., 1994. The plesiosaur's birthplace: the Bristol Institution and its contribution to vertebrate palaeontology. *Zoological Journal of the Linnean Society*, v.112, 179-196; and Evans, M., 2010. The roles played by museums, collections and collectors in the early history of reptile palaeontology. *In* Moody, R.T.J., Buffetaut, E., Naish, D. & Martill, D.M. (eds), *Dinosaurs and Other Extinct Saurians: a Historical Perspective*. The Geological Society, London, Special Publications, No.343, 17.

36. North, F.J., 1933. Dean Conybeare, Geologist. *Transactions of the Cardiff Naturalists' Society*, v.66, 31.

37. Buckland's description of the plesiosaur is from Buckland, W., 1836. *Geology and Mineralogy considered with reference to Natural Theology*, London: William Pickering. v.1, 202.

38. Conybeare included a similar description of the possible mode of life of *Plesiosaurus* in his published paper: Conybeare, W.D., 1824. On the discovery of an almost perfect skeleton of a Plesiosaurus. Read February 20, 1824. *Transactions of the Geological Society of London, Second Series*, v.1, 380-387. It was also published in the *Philosophical Magazine*, v.65, 412-421.

39. William Wyndham Grenville, 1st Baron Grenville (1759-1834) joined the Geological Society in 1821; his patronage of Buckland was influential in establishing Buckland's post as Reader in Geology. Letter Buckland to Lord Grenville 25 February 1824 Dropmore papers Add 58995 ff.109 & 110. I am grateful to Peter Lincoln for drawing this letter to my attention. Buckland had taken over as President at the Society's Anniversary Meeting on 6 February.

40. On Wednesday 5 May 1824, the *Hereford Journal* reported that 'the great Cuvier, having received a sketch of the fossil discovered at Lyme, wrote to London to state his opinion that the head could not have belonged to so small a body, it being well known to geologists that fragments of different animals are frequently found lying immediately contiguous ... the contents of the blue lias at

Lyme are so numerous and extraordinary, that he shall not be surprised at any discovery that may be made there.' What this report probably meant to say was that the head seemed too small for the body. Anning is acknowledged on page 473 of Cuvier, G., 1824. *Recherches sur les ossemens fossiles, où l'on rétablit les caractères de plusieurs animaux dont les révolutions du globe ont détruit les espèces*. 2nd edition. Tome cinquième, Deuxième partie, contenant les ossemens de reptiles et le résumé général. Paris & Amsterdam: Dufour & d'Ocagne.

41. Constant Prévost (1787-1856). Lyell's letter of 9 July 1824 to Mantell is quoted in Lyell, Mrs [K.M.] (ed.), 1881. *Life letters and journals of Sir Charles Lyell, Bart*. 2 vols. London: John Murray, 153.

42. Henry Waring RN (c.1773-1837). The report on the plesiosaur discovery by Waring is from *The New Monthly Magazine and Literary Journal*, v.12, 1824, 1 February, 92-93. For more on the specimen, sold to Anning and bought by Prévost, see Taquet, P., 2003. Quand les reptiles marins anglais traversaient la manch. Mary Anning et Georges Cuvier, deux acteurs de la découverte et de l'étude des ichthyosaures et des plésiosaures. *Annales de Paléontologie*, v.89, 37-64; and Vincent, P., Taquet, P., Fischer, V., Bardet, N., Falconnet, J. & Godefroit, P., 2014. Mary Anning's legacy to French vertebrate palaeontology, *Geological Magazine*, v.151, 7-20.

43. Prévost, C., 1825. Note sur le gisement des ossemens fossiles d'Ichthyosaures et de Plésiosaures dans les couches du Lias de Lyme Regis. *Nouveau Bulletin des Sciences, par la Société Philomatique de Paris*, 1825, 167-170.

44. Thomas Allan (1777-1833). Allan's view of Anning is from his notebook, *Travels in England 1813-24*, in the Library of the Natural History Museum and is quoted by Lang, W.D., 1939. Mary Anning (1799-1847) and the pioneer geologists of Lyme. *Proceedings of the Dorset Natural History & Archaeological Society*, v.60, 153-154.

45. Harriet Silvester (1753-1843). Lady Silvester's comments are from Welch, E., 1967. Lady Silvester's tour through Devonshire in 1824. *Devon and Cornwall Notes and Queries*, v.30, 313 and Welch, E., 1973. Lady Silvester's tour through Devonshire in 1824. *Devon and Cornwall Notes and Queries*, v.32. 265-266.

46. Barber, L., 1980. *The heyday of natural history 1820-1870*. London: Jonathan Cape, 127, describes Anning as 'barely literate'. The printed copy of Conybeare's 1824 paper from the *Transactions of the Geological Society of London* which he presented to Elizabeth Philpot is in the collection of Lyme Regis Museum. I am grateful to Richard Bull for bringing it to my attention.

47. Anning's hand-written copies of Conybeare's papers are in the Natural History Museum Library. As her friendship with Conybeare endured throughout her life, she seems not to have taken offence; perhaps she put it down to Conybeare's shy and reserved nature.

48. Fanny's correspondence with her friends, including Mary Anning, is

quoted from Grant, J., 1827. *A memoir of Miss Frances Augusta Bell, who died in Kentish Town, on Monday, the 23rd of May, 1825, aged fifteen years and six months: with specimens of her compositions, in prose and verse.* London: Hatchard & Son.

49. Andrew Ure (1778-1857), *A New System of Geology, in which the Great Revolutions of the Earth and Animated Nature, are reconciled at once to Modern Science and Sacred History.* London: Longman, Rees, Orme, Brown, & Green, 228-237.

50. Joseph Anning's letter to Charles Konig, 13 September 1825, Natural History Museum Library, DF100/1/1.

51. Mary Anning to Charlotte Murchison 11 October [1833], Geological Society Archives, LDGSL/838/A/7/3. Quoted in Lang, W.D., 1945. Three letters by Mary Anning, Fossilist of Lyme. *Proceedings of the Dorset Natural History & Archaeological Society*, v.66, 169-173.

52. Reverend Johnson Grant (1773-1844), 1827. *A memoir of Miss Frances Augusta Bell, who died in Kentish Town, on Monday, the 23rd of May, 1825, aged fifteen years and six months: with specimens of her compositions, in prose and verse.* London: Hatchard & Son. Fanny's friend Anne is probably Mary Anne Davis whose *Tributary stanzas to the memory of a young friend of extraordinary talents and virtues* was published in Grant's book. She was also the author of *Fables in verse: from Aesop. La Fontaine and others*, London, 1813, and *A selection from the parables of the New Testament, paraphrased in familiar verse for the use of children*, published in Frome in 1836, as well as a number of other devotional texts and poems.

53. Watney, M. 1931. Mary Anning – first woman geologist. *Morning Post*, 4 February 1931, 8; and Lang, W.D., 1939 Mary Anning (1799–1847), and the pioneer geologists of Lyme, *Proceedings of the Dorset Natural History and Archaeological Society*, v.60, 144 state that the Anning house was damaged in the Great Storm of 1824. Reports of the storm can be found in Roberts, G. 1824. *An accurate account of the storm at lyme, on the 23rd of November, 1824. As contained in a Letter addressed to the Editor of the Sherborne Mercury, with a supplement, containing remarks up to the present time.* Lyme Regis: Jessep; Roberts, G., 1834. *The History and Antiquities of the Borough of Lyme Regis and Charmouth.* London: Samuel Bagster; Le Pard, G. 2000. The Great Storm of 1824. *Proceedings of the Dorset Natural History and Archaeological Society*, v.121, 23–36; and Taylor, M.A., 2020. Mary Anning of Lyme Regis, and the Great Storm of 1824. *Proceedings of the Dorset Natural History and Archaeological Society*, v.141, 31–40.

54. Conybeare's visit to Lyme, based on local press reports and the Annual Reports of the Bristol Institution is described by Taylor, M.A. & Torrens, H.S., 1986. Saleswoman to a new science: Mary Anning and the fossil fish *Squaloraja*

from the Lias of Lyme Regis. *Proceedings of the Dorset Natural History and Archaeological Society*, v.108, 135-148.

55. De la Beche, H.T., 1826. Notice of Traces of a Submarine Forest at Charmouth, Dorset. *Annals of Philosophy*, New Series, v.11, February 1826, 143. The discovery was reported in the *Bath Chronicle and Weekly Gazette*, Thursday 27 October 1825, 2.

56. The description of Anning's house is from *Lyme 50 years ago*, published in 1892, and quoted by Draper, J., 2004. *Mary Anning's town – Lyme Regis*. Lyme Regis: Lyme Regis Museum.

6. DECADE OF DISCOVERIES

1. George Brettingham Sowerby (1788-1854); James Sowerby (1757-1822). Mary Anning to G.B. Sowerby 27 July 1826. G.B. Sowerby correspondence no 329a. Anning's letters to G.B. Sowerby are in the G.B. Sowerby Archive in the National Museum of Wales. See Matheson, C., 1964. George Brettingham Sowerby the First and his correspondents. *Journal of the Society for the Bibliography of Natural History*, v.4, 214-225.

2. Thompson, F.M.L., 2004. Grenville, Richard Temple-Nugent-Brydges-Chandos-, first Duke of Buckingham and Chandos (1776–1839). *Oxford Dictionary of National Biography*, describes Grenville as 'odious and unpopular ... utterly without talent' while Sack, J.J., 1975. The decline of the Grenvillite Faction under the First Duke of Buckingham and Chandos, 1817-1829. *Journal of British Studies*, v.15, 112-134 elaborates on the 'vain, socially foolish, politically irresponsible' Grenville. For details of the acquisition of the ichthyosaur by the Bristol Institution see Taylor, M.A. & Torrens, H.S., 1987. Saleswoman to a new science: Mary Anning and the fossil fish *Squaloraja* from the Lias of Lyme Regis. *Proceedings of the Dorset Natural History and Archaeological Society*, v.108, 139; and Taylor, M.A., 1994. The plesiosaur's birthplace: the Bristol Institution and its contribution to vertebrate palaeontology. *Zoological Journal of the Linnean Society*, v.112, 179-196. The specimen, registration number Cb 2465, was lost when Bristol City Museum, to which the Institution's specimens were transferred, was bombed during the Bristol Blitz of November 1940.

3. Mary Anning to G.B. Sowerby 4 October 1826. G.B, Sowerby correspondence no 342, National Museum of Wales.

4. Adam Sedgwick to G.B. Sowerby 30 October 1826. G.B. Sowerby correspondence no 562, National Museum of Wales.

5. Mary Anning to G.B. Sowerby 3 March 1829. G.B. Sowerby correspondence no 414, National Museum of Wales.

6. Mary Anning to G.B. Sowerby 9 May 1829. G.B. Sowerby correspondence no 488, National Museum of Wales.

7. Mary Anning to G.B. Sowerby 15 May 1829. G.B. Sowerby correspondence

NOTES, SOURCES AND FURTHER READING

no 795, National Museum of Wales.

8. George William Featherstonhaugh (1780-1866). Featherstonhaugh's visit to England and Lyme is from Berkeley, E. & Berkeley, D.S., 1988. *George William Featherstonhaugh. The First U.S. Government Geologist*. Tuscaloosa and London: University of Alabama Press.

9. Johann Friedrich Blumenbach (1752-1840) named the crinoid *Pentacrinites fossilis* for the five-sided segments which made up the crinoid's stem. It was later named *Pentacrinites briareus* or called the 'Briarian Pentacrinite' by palaeontologists in England, and this is the name often used by Anning. One particular rock layer on Black Ven is known as the 'Pentacrinite Bed', although the crinoid also occurs within the marls just below it. See Simms, M.J., 1989. *British Lower Jurassic Crinoids*. Monograph of the Palaeontographical Society Publication No. 581. London: The Palaeontographical Society, for everything you ever wanted to know about *Pentacrinites*. The sale prices of these fossils come from Mantell, G.A., 1846. A few Notes on the Prices of Fossils. *London Geological Journal*, v.1, 14; and from Rolfe, W.D.I., Milner, A.C. & Hay, F.G., 1988. The price of fossils. *In* Crowther, P.R. & Wimbledon, W.A. (eds) *Special Papers in Palaeontology*, no.40, 139-171.

10. The loss of the Lyceum's collections is noted in Barnhart, J.H., 1917. The first hundred years of the New York Academy of Sciences. *The Scientific Monthly*, v.5, 463-475.

11. *Bristol Mirror*, Saturday 29 December 1827, 3. Also *Bristol Mercury*, Monday 31 December 1827, 3; *Exeter Flying Post*, Thursday 27 December 1827, 4; and others.

12. The discovery of this *Dapedium* was reported in the *Globe*, Tuesday 25 November 1828 and across the country in the *London Evening Standard* Thursday 27 November 1828; the *Sherborne Mercury*, *Bristol Mirror* and *Yorkshire Gazette* Saturday 29 November 1828; *Chester Courant* Tuesday 2 December 1828; *North Wales Chronicle* Thursday 4 December 1828.

13. Mary Anning to William Buckland 24 November 1834, private collection. The belemnite-rich beds she describes at Golden Cap are probably those known today as the Belemnite Stone and the Belemnite Bed which occur at the top of the Belemnite Marl Member of the Charmouth Mudstone Formation. Her observations are included by Buckland on page 377 of his *Bridgewater Treatise* of 1836.

14. Sir Francis Chantrey (1781-1841).

15. Elizabeth Philpot to Mary Buckland, 9 December 1833, Oxford University Museum of Natural History. Philpot made several sepia drawings of this skull; another she sent or gave to Henry De la Beche who presented his to the Geological Society in March 1834. Both the Oxford and Geological Society drawings survive.

16. William Buckland and Mary Buckland to Elizabeth Philpot, 1 January

1834, private collection. Elizabeth Philpot to William Buckland [January 1834] Oxford University Museum of Natural History.

17. The fossil sepia named by Buckland is now called *Loligosepia bucklandi*. See Donovan, D.T. & von Boletzky, S., 2014. *Loligosepia* (Cephalopoda: Coleoidea) from the Lower Jurassic of the Dorset coast, England. *Neues Jahbuch für Geologie und Paläontologie*, v.273/1, 45-63.

18. Mary Anning's dissection is in her letter to William Buckland 21 December 1830 and is quoted by Edmonds, J.M., 1978. The fossil collection of the Misses Philpot of Lyme Regis. *Proceedings of the Dorset Natural History & Archaeological Society*, v.98, 43-48; and in Lang, W.D., 1939. Mary Anning (1799-1847) and the pioneer geologists of Lyme. *Proceedings of the Dorset Natural History & Archaeological Society*, v.60, 142-164.

19. *The New Monthly Magazine and Literary Journal*, 1 July 1828, 323.

20. Buckland, W., 1835. On the discovery of coprolites, or fossil faeces, in the Lias at Lyme Regis, and in other formations. Read 6th February 1829. *Transactions of the Geological Society of London, Second Series*, v.3, 223-236.

21. William Wollaston (1766-1828); William Prout (1785-1850). That Anning had recognised coprolites as fossil faeces as early as 1824 is indicated in correspondence between Roderick Murchison and George Featherstonhaugh on 13 May and 20 June 1829. See Berkeley, E. & Berkeley, D.S., 1988. *George William Featherstonhaugh. The First U.S. Government Geologist*. Tuscaloosa and London: University of Alabama Press.

22. Buckland, W., 1823. *Reliquiae Diluvianae; or, Observations on the Organic Remains contained in Caves, Fissures, and Diluvial Gravel, and on other geological phenomena, attesting the action of a Universal Deluge*. London: John Murray, 37-38.

23. The two lines are from a longer poem sent by Philip Bury Duncan (1772-1863) to Buckland. See Duffin, C.J., 2009. 'Records of warfare ... embalmed in the everlasting hills': a history of early coprolite research. *Mercian Geologist*, v.17, 108.

24. Buckland's coprolite table is described by Bull, R., 2010. *William Buckland's coprolite table. Objects in the Museum Paper* 1/2010 which can be found via the website of Lyme Regis Museum. The quote about the table being in the drawing room is from Buckland, F., 1867. *Curiosities of natural history*. London: Richard Bentley.

25. For modern accounts of coprolites and Buckland's work, see Duffin, C.J., 2009. 'Records of warfare ... embalmed in the everlasting hills': a history of early coprolite research. *Mercian Geologist*, v.17, 101-111; Duffin, C.J., 2012. Coprolites and characters in Victorian Britain. *In* Hunt, A.P., Milàn, J., Lucas, S.G. & Speilmann, J.A. (eds), Vertebrate coprolites. *New Mexico Museum of Natural History and Science, Bulletin* 57, 45-60; and Pemberton, S.G., 2012.

William Buckland (1784-1856) and Henry De la Beche (1796-1855): the early history of coprolites. *In* Hunt, A.P., Milàn, J., Lucas, S.G. & Speilmann, J.A. (eds), Vertebrate coprolites. *New Mexico Museum of Natural History and Science, Bulletin* 57, 29-44.

26. Buckland, W., 1836. *Geology and Mineralogy considered with reference to Natural Theology.* London: William Pickering, vol.1, 201-202.

27. Buckland, W., 1829. On the discovery of a new species of Pterodactyle; also of the faeces of the Ichthyosaurus; and of a black substance resembling sepia, or Indian ink, in the Lias at Lyme Regis. *Proceedings of the Geological Society of London*, v.1, No.10, 96.

28. William Rowe (c.1747-1810). For more on the history of the discovery of pterosaurs see Wellnhofer, P., 2008. A short history of pterosaur research. *Zitteliana*, B28, 7-19; Martill, D.M., 2010. The early history of pterosaur discovery in Great Britain. *In* Moody, R.T.J., Buffeteaut, E., Naish, D. & Martill, D.M. (eds), *Dinosaurs and other extinct saurians: a historical perspective.* The Geological Society, London, Special Publications, No.343, 287-311. The pterosaur jaw fragment seen by Buckland in the Philpot collection and illustrated in his paper is now in Oxford University Museum of Natural History, OUM J 28251.

29. Buckland, W., 1835. On the discovery of a new species of Pterodactyle in the Lias at Lyme Regis, and in other formations. Read 6th February 1829. *Transactions of the Geological Society of London, Second Series*, v.3, 217-218.

30. See Martill, D.M., 2014. *Dimorphodon* and the Reverend George Howman's noctivagous flying dragon: the earliest restoration of a pterosaur in its natural habitat. *Proceedings of the Geologists' Association*, v.125, 120-130.

31. Reverend George Ernest Howman (1797-1878) had been a student at Balliol College, Oxford from 1815 to 1821 and was vicar at Sonning from 1822 to 1841. Lady Isobel Kerr (1881-1975); John Kerr, 7th Marquis of Lothian (1794-1841). That Howman had an interest in fossils is shown by a letter of 1837 to the *Magazine of Natural History*, v.1, New Series, 532: see Knell, S.J., 2000. *The culture of English Geology, 1815-1851. A science revealed through its collecting.* Aldershot: Ashgate, 218.

32. Lyell, Mrs [K.M.] (ed.), 1881. *Life letters and journals of Sir Charles Lyell, Bart.* London: John Murray, vol. 1, 247.

33. Anning's 1828 pterosaur is in the collections of the Natural History Museum, registration number NHMUK R 1034. Owen's naming of the specimen is in Owen, R., 1859. On a new genus (*Dimorphodon*) of pterodactyl, with remarks on the geological distribution of flying reptiles. *Report of the British Association for the Advancement of Science*, 1858, 97-98. More details can be found in Owen, R., 1870. *A monograph of the fossil Reptilia of the Liassic formations, Part 3*, Dimorphodon. *Monograph of the Palaeontographical Society*, 23, 41-82.

34. Mary Anning to William Buckland, 24 November 1834, private collection. Buckland, W., 1836. *Geology and Mineralogy considered with reference to Natural Theology.* Vol 1. London: William Pickering, 437-439.

35. The discovery of the complete *Plesiosaurus dolichodeirus* in 1829 was reported in the *Dorset County Chronicle*, Thursday 19 February 1829, 4; and in *The Crypt and West of England Magazine*, New Series No.4, April 1829, 164. Mary Anning's letter to William Buckland of 9 February 1829 is in the Natural History Museum Library Archives DF100/2/5.

36. Mary Anning to William Buckland, 14 February 1829, Natural History Museum Library Archives DF100/2/6.

37. William Buckland to Charles Konig, 23 February 1829, Natural History Museum Library Archives DF100/2/4.

38. Draft copy of letter from Charles Konig to Mary Anning 1 April 1829, Natural History Museum Library Archives DF100/2/2.

39. The report of its purchase by the British Museum is in the *Exeter Flying Post*, Thursday 7 May 1829, 2-3. The specimen is still in the Natural History Museum, registration number NHMUK R 1313.

40. Geikie. A., 1875. *Life of Sir Roderick I. Murchison.* London: John Murray, vol. 2, 334.

41. Charlotte's parents were General Francis Lewis Hugonin (1750-1836) and Charlotte Edgar (*died* 1838).

42. Mary Somerville (1780-1872) became a lifelong friend of the Murchisons and these quotes from her 1874 autobiography are taken from Kölbl-Ebert, M., 1997. Charlotte Murchison (née Hugonin) 1788-1869. *Earth Sciences History*, v.16, 39-40. For more on Roderick and Charlotte Murchison, see Bonney, T.G., 2004. Murchison, Sir Roderick Impey, baronet (1792-1871). *Oxford Dictionary of National Biography*; Kölbl-Ebert, M., 2004. Murchison [née Hugonin], Charlotte, Lady Murchison (1788-1869) *Oxford Dictionary of National Biography;* Bassett, D.A., 1991. Roderick Murchison's The Silurian System: a sesquicentennial tribute. *In* Bassett, M.G., Lane, P.D. & Edwards, D. (eds), *The Murchison Symposium. Proceedings of an International Conference on the Silurian System.* London: The Palaeontological Association, 7-90.

43. The term 'Silurian System' refers to the rocks deposited during the Silurian Period of geological time; similarly for the Devonian System and Permian Systems which refer to rocks of the Devonian Period and Permian Period. The Silurian Period extends from 443 million years ago to 419 million; the Devonian from 419 million to 359 million years ago; and the Permian from 299 million to 252 million years ago. The Jurassic Period on whose rocks Mary Anning worked, extends from 201 million to about 145 million years ago. These dates can be found on the *International Chronostratigraphic Chart* v.2020/03 published by the International Commission on Stratigraphy, part of the International Union of

Geological Sciences.

44. Mary Anning's letters to Charlotte Murchison are in the Geological Society Archives, LDGSL/838/A/7/1-3. They are quoted in full in Lang, W.D., 1945. Three letters by Mary Anning, Fossilist of Lyme. *Proceedings of the Dorset Natural History & Archaeological Society*, v.66, 169-173.

45. William Lonsdale (1794-1871). Anning's notes on her London visit are in the Library of the Natural History Museum. Robert Edmond Grant (1793-1874), appointed to the post in 1827, built up a zoological teaching collection which numbered about 6,000 specimens by 1829. This collection is the foundation of the Grant Museum of Zoology, University College London. I am grateful to Wendy Cawthorne of the Geological Society for her help trying to identify the 'zoological rooms' visited by Anning and to Tannis Davidson, Curator of the Grant Museum of Zoology.

46. St Mary's, Bryanston Square was built 1823-24; Thomas Frognall Dibdin (1776-1847) was rector from 1824 until his death.

47. The papers Anning refers to here are probably Murchison, R., 1826. Geological sketch of the north-western extremity of Sussex and the adjoining parts of Hants and Surrey. *Transactions of the Geological Society of London*, Series 2, v.2, 97-107; De la Beche, H.T., 1826. On the Chalk and the sands beneath it (usually termed Green-sand) in the vicinity of Lyme Regis, Dorset and Beer, Devon. *Transactions of the Geological Society of London*, Series 2, v.2, 109-118; and Fitton, W.H., 1824. Inquiries respecting the geological relations of the beds between the Chalk and the Purbeck Limestone on the South-east of England. *Annals of Philosophy*, v.24, New Series 8, 365-458. For the geologically-inclined, a summary of the disagreements on the Greensand strata can be found in Jukes-Browne, A.J., 1900. *The Cretaceous rocks of Britain. Vol 1. The Gault and Upper Greensand of England. Memoirs of the Geological Survey.* London: HMSO, 15-26.

48. Owen describes his visit to Lyme and being caught by the tide in a letter of 13 September 1839 to William Clift which is quoted in Owen, R., 1894. *The life of Richard Owen*. vol. 1. London: John Murray, 164-166.

49. For more on *A Coprolitic Vision*, see McCartney, P.J., 1977. *Henry De la Beche: observations on an observer.* Cardiff: Friends of the National Museum of Wales, 48-49.

7. MARY AND THE SEA DRAGONS

1. *Salisbury and Winchester Journal*, Monday 28 December 1829, 2; *Manchester Mercury*, Tuesday 5 January 1830, 1; *Yorkshire Gazette*, Saturday 2 January 1830, 3.

2. Mary Anning to J.S. Miller, 30 January [1830], quoted in Taylor, M.A. & Torrens, H.S., 1987. Saleswoman to a new science: Mary Anning and the fossil

fish *Squaloraja* from the Lias of Lyme Regis. *Proceedings of the Dorset Natural History and Archaeological Society*, v.108, 135. This paper gives a full account of Anning's December 1829 discovery of the new fossil fish and its acquisition by the Bristol Institution.

3. Mary Anning to William Buckland 15 February [1830], Sotheby's sale catalogue *Books and Manuscripts: A Summer Miscellany*, 28 July 2020.

4. Lyell's letter to Mantell of 23 April 1830 is published in Lyell, Mrs [K.M.] (ed.), 1881. *Life letters and journals of Sir Charles Lyell, Bart*. London: John Murray. vol. 1, 265.

5. Delays in the negotiations with the Bristol Institution may have been caused by the death of the curator, Miller, on 25 May 1830. The letter from Mary Anning to Adam Sedgwick is dated 11 February [1831] is quoted from Taylor, M.A. & Torrens, H.S., 1987, 136 (see note 2, above). A copy of Anning's drawing of *Squaloraja* is in the Buckland Papers at Oxford University Museum of Natural History. See Kölbl-Ebert, M., 2012. Sketching rocks and landscape: drawing as a female accomplishment in the service of geology, *Earth Sciences History*, v.31, 270-286.

6. John Naish Sanders (c.1777-1870).

7. Henry Riley (1797-1848). Riley, H., 1833. On a fossil in the Bristol Museum and discovered in the Lias of Lyme Regis. *Proceedings of the Geological Society of London*, v.1, 483-484; Riley, H., 1837. On Squaloraia. *Transactions of the Geological Society of London*, Series 2, v.5, 83-88. For more on Riley, see Taylor, M.A. & Torrens, H.S., 2017. Lost & Found. 280. Henry Riley M.D. (1797-1848) of Bristol. *The Geological Curator*, v.10, 493-498.

8. The tale of *Squaloraja*'s tail is described by Taylor & Torrens 1987 (see note 2, above), and by Edmonds, J.M., 1978. The fossil collection of the Misses Philpot of Lyme Regis. *Proceedings of the Dorset Natural History & Archaeological Society*, v.98, 43-48 who quotes the letter from Sanders to Elizabeth Philpot which is dated 10 February 1838. The tail is Oxford University Museum of Natural History, registration number OUM J3097.

9. The original watercolour of De la Beche's lithograph was rediscovered in 1980 in a London art dealer's window display, and was acquired by the Department of Geology at the National Museum of Wales for its De la Beche Archive where its accession number is NMW84.20G.D368.

10. The price of the *Duria* lithograph would have been the equivalent of over four week's wages for a literate lower middle class worker such as a lawyer's clerk; see O'Connor, R., 2007. *The Earth on show. Fossils and the poetics of popular science, 1802-1856*. Chicago and London: University of Chicago Press, 219-223. Letter Charles Lyell to Gideon Mantell 13 May 1830, Mantell MSS, Alexander Turnbull Library, Wellington, New Zealand. Lyell may have been less enamoured of another sketch, *Awful Changes*, drawn by De la Beche in 1830

criticising Lyell's cyclical ideas of earth history. For more see McCartney, P.J., 1977. *Henry De la Beche: observations on an observer.* Cardiff: Friends of the National Museum of Wales, 50-53.

11. Buckland's delight is expressed in his letters to De la Beche on 25 May [1830] and 1 May 1831 in the De la Beche Archive, in the Department of Geology, National Museum of Wales, NMW 84.20G.D182 and 180. Several different versions of the lithograph were printed, some hand-coloured, and copies are rare. That he sent 100 copies to Anning is mentioned in a draft of a letter to De la Beche on 14 October 1831 but was not included in the letter which was sent; Buckland Papers in Oxford University Museum of Natural History.

12. For more on *Duria antiquior*, see McCartney, P.J., 1977. *Henry De la Beche: observations on an observer.* Cardiff: Friends of the National Museum of Wales; Rudwick, M.J.S., 1992. *Scenes from deep time. Early pictorial representations of the prehistoric world.* Chicago & London: University of Chicago Press; and Sharpe, T. & Clary, R.M., 2021. Henry De la Beche's 1830 pioneering paleoecological illustration, *Duria antiquior.* In Clary, R.M., Rosenberg, G.D. & Evans, D. (eds), *The Evolution of Paleontological Art.* Geological Society of America Memoir 218, 1-8.

13. Buckland's letter to De la Beche 15 September 1830 is in the De la Beche Archive, Department of Geology, National Museum of Wales, NMW84.20G. D179.

14. Mary Anning to Charlotte Murchison, 11 October 1833, Geological Society Archives, LDGSL/838/A/7/3.

15. Anning's letter to Buckland, 21 December 1830 is quoted by Lang, W.D., 1939. Mary Anning (1799-1847) and the pioneer geologists of Lyme. *Proceedings of the Dorset Natural History & Archaeological Society,* v.60, 142-164.

16. Anning's letter to Sedgwick is dated 11 February [1831] and is quoted from Taylor, M.A. & Torrens, H.S., 1987. Saleswoman to a new science: Mary Anning and the fossil fish *Squaloraja* from the Lias of Lyme Regis. *Proceedings of the Dorset Natural History and Archaeological Society,* v.108, 136. The original is in Cambridge University Library, CUL Add. MS 7652.

17. William Willoughby Cole (1807-1886). For more on Lord Cole and his fossil fish collection see James, K.M., 1986. *"Damned nonsense!" The geological career of the third Earl of Enniskillen.* Belfast: Ulster Museum. Philip de Malpas Grey Egerton (1806-1881).

18. The plesiosaur was named in Buckland, W., 1836. *Geology and Mineralogy considered with reference to Natural Theology.* 2 vols. London: William Pickering. It was described by Owen, R., 1840. A description of a specimen of the *Plesiosaurus macrocephalus,* Conybeare in the collection of Viscount Cole, MP, DCL, FGS &c. *Transactions of the Geological Society of London, Second Series,* v.5, 515-535. The specimen is in the Natural History Museum, registration

number NHM UK R1336.

19. William John Broderip (1789-1859). Broderip, W.J., 1837.Description of some Fossil Crustacea and Radiata, found at Lyme Regis, in Dorsetshire. Read June 10, 1835. *Transactions of the Geological Society of London, Second Series*, v.5, 171-174.

20. For more on the specimens Sedgwick purchased from Anning see Price, D., 1986. Mary Anning specimens in the Sedgwick Museum, Cambridge. *The Geological Curator*, v.4, 319-324. The specimens were identified by the late David Price in the Sedgwick Museum collections as J.59645, the ichthyosaur skull; and J.59642, the skeleton missing its skull. Price noted that both are set in mortar in a shallow wooden black-painted frame.

21. Buckland's letter of 1 May 1831 to De la Beche is in the De la Beche Archive at the National Museum of Wales, NMW84.20G.D180. Beriah Botfield (1807-1863) and William Devonshire Saull (1783-1855) were Fellows of the Geological Society.

22. Sir Astley Paston Cooper (1768-1841); Anning's letter to Cooper is undated and is in The Women's Library Special Collections 9/02/003, London School of Economics. Lyell's letter of 10 April 1832 to Gideon Mantell is quoted from Wennerbom, A.J., 1999. *Charles Lyell and Gideon Mantell, 1821-1852: their quest for elite status in English geology*. Unpublished PhD thesis, University of Sydney; Sir Astley Cooper's visit to Lyme and his remarks about Anning are from Cooper, B.B., 1843. *The life of Sir Astley Cooper, Bart., interspersed with sketches from his note-books of distinguished contemporary characters*, v.2, 385-386. London: John W. Parker. For more on Cooper's ichthyosaur, now identified as *Ichthyosaurus breviceps*, see Taylor, M.A., 2014. Rediscovery of an *Ichthyosaurus breviceps* Owen, 1881 sold by Mary Anning (1799-1847) to the surgeon Astley Cooper (1768-1841) and figured by William Buckland (1784-1856) in his *Bridgewater Treatise. Geoscience in South-West England*, v.13, 321-327. The specimen is in the Natural History Museum, registration no. NHMUK PV R.8437.

23. The mention of the specimen collected by Anning in Watchet in Somerset is in Buckland, W., 1836. *Geology and Mineralogy considered with reference to Natural Theology*. London: William Pickering. Volume 2, 51. The specimen is catalogued as *Belemnites* sp. ink sac, registration number OUMNH J.03564.

24. The fossil fish identified by Agassiz were from a distinctive bed of rock containing fish scales, bones, teeth and coprolites known as the Rhaetic Bone Bed. It occurs at the base of the Triassic Westbury Formation. Sir Philip Egerton describes it as 'a thin stratum replete with remains of saurians and ichthyolites ... near Axmouth' and refers to the specimens collected by Anning. See Egerton. P.G., 1841. A Notice on the Occurrence of Triassic Fishes in British Strata. *Proceedings of the Geological Society of London*, v.3, no.77, 409-410. A letter from Elizabeth

Philpot to William Buckland 26 June 1835 (private collection) tells us that Anning was collecting near Bridport. The specimens were most likely *Balanocrinus gracilis* from the Down Cliff Sand Member of the Dyrham Formation, a younger part of the Lower Jurassic sequence than that from which she usually collected. I am grateful to Mike Simms for his assistance in identifying the crinoid and its probable locality. Anning's finances were again precarious by the mid 1830s, which would likely rule out any paid coach travel or accommodation.

25. Susannah-Catherine, Lady Duckworth (*d.*1840); John Tomas Duckworth (1748-1817); part of Lady Paterson's diary, which is preserved in Devon County Record Office, is published in Hunt, P. (ed.), 1984. *Devon's Age of Elegance. Described by the diaries of the Reverend John Swete, Lady Paterson and Miss Mary Cornish*. Exeter: Devon Books.

26. The letter from Buckland to Joseph Anning is in the Department of Geology National Museum of Wales NMW00.16G.D2.

27. Reverend David Williams (1792–1850). The account of Mantell's visit to Lyme is in Curwen, E.C., 1940. *The journal of Gideon Mantell surgeon and geologist covering the years 1818-1852*. London: Oxford University Press.

28. Mantell's view of Anning may have been coloured by his personal marital difficulties at this time, brought about by his preoccupation with building his fossil collection and making a name for himself amongst his geological peers. Mantell and his wife eventually separated in 1839. The suggestion that Mantell may have confused Molly Anning with her daughter comes from Goodhue, T.W., 2004. *Fossil hunter. The life and times of Mary Anning (1799-1847)*. Bethesda, Maryland: Academic Press, 77.

29. Thomas Hawkins (1810-1889). Lang's description of Thomas Hawkins is in Lang, W.D., 1939. Mary Anning (1799-1847) and the pioneer geologists of Lyme. *Proceedings of the Dorset Natural History & Archaeological Society*, v.60, 142-164; see also Taylor, M.A., 2000. Mary Anning, Thomas Hawkins and Hugh Miller, and the realities of being a provincial fossil collector. *The Edinburgh Geologist*, v.34, 28-37; and Taylor, M.A., 2004. Hawkins, Thomas (1810-1889). *Oxford Dictionary of National Biography*. More on Hawkins can be found in Bulleid, A., 1843. Notes on the life and work of Thomas Hawkins, F.G.S. *Proceedings of the Somerset Archaeological and Natural History Society*, v.89, 59-71. Mantell's description is in a letter of 18 June 1834 to the American chemist Benjamin Silliman (1779-1864), transcript with the De la Beche Archive, National Museum of Wales.

30. This is likely the discovery reported in the *Morning Advertiser*, Saturday 1 September 1832, 4: 'Miss M.Anning lately discovered a stupendous fossil animal high up in the cliffs east of Lyme, and sold it, in its imbedded state, to Mr. Hawkins, an amateur geologist who went to the expense of extricating it, and having it conveyed to his collection'.

31. Hawkins' description of his July 1832 visit to Lyme and excavation of the large ichthyosaur and of his 1833 visit to Charmouth is from Hawkins, T., 1834. *Memoirs of Ichthyosauri and Plesiosauri, extinct monsters of an ancient Earth*. London: Rolfe & Fletcher, as are the subsequent quotes. The specimen, *Temnodontosaurus platyodon*, is in the Natural History Museum, registration no. NHMUK 2003. Evidence for likely *Temnodontosaurus* predation on another large ichthyosaur was uncovered by fossil-hunter Chris Moore of Charmouth in the winter of 2015-16; see *Attenborough and the Great Sea Dragon* by Chris Moore, self-published c.2018.

32. Murchison, R.I., 1833. Presidential Address. *Proceedings of the Geological Society of London*, v.1, No.30, 445.

33. The 11 October 1833 letter with Mary Anning's comment on Roderick Murchison is in the Geological Society Archives, LDGSL/838/A/7/3. She seems to have found rugged men attractive; gossiping with Anna Pinney she called 'young dandies ... numskulls, not men'.

34. Conybeare's letter of 4 August 1834 to Buckland in in the Department of Geology, National Museum of Wales, NMW84.20G.D309. Mantell's comment was noted in his journal on Tuesday 10 June 1834: 'Received Mr Hawkins' strange but splendid book on the Plesiosauri'. See Curwen, E.C., 1940. *The journal of Gideon Mantell surgeon and geologist covering the years 1818-1852*. London: Oxford University Press.

35. Hawkins 1840 book is *The book of the great sea dragons, Ichthyosauri and Plesiosauri, gedolim taninim, of Moses. Extinct monsters of the ancient earth*. London: William Pickering. The late nineteenth century author who was dismissive of Hawkins' books was Henry Neville Hutchinson (1856-1927) in his 1893 book *Extinct Monsters: a popular account of some of the larger forms of ancient animal life*. London; Chapman & Hall, 39. Stephen Jay Gould's quote is from Purcell, R.W. & Gould, S.J., 1992. *Finders, keepers. Eight collectors*. London: Hutchinson Radius, 107. However, for a view of Hawkins' writings within the literary context of the period, see O'Connor, R., 2003. Thomas Hawkins and geological spectacle. *Proceedings of the Geologists' Association*, v.114, 227-241; and Carroll, V., 2007. "Beyond the Pale of Ordinary Criticism". Eccentricity and the Fossil Books of Thomas Hawkins. *Isis*, v.98, 225-265.

36. John Samuelson Templeton *(fl.*1830-1857); John Martin (1789-1854). For an interpretation of Hawkins' writings in the literary context of the time see O'Conner, R., 2003. Thomas Hawkins and geological spectacle. *Proceedings of the Geologists' Association*, v.114, 227-241 and O'Connor, R., 2007. *The Earth on show. Fossils and the poetics of popular science, 1802-1856*. Chicago and London: University of Chicago Press.

37. The coprolites collected by Anning and in the Hawkins' collection in the Natural History Museum are NHMUK 2066 and 2066a. See Chapman, S. &

Milner, A.C., 2010. Appendix. A catalogue of fossil specimens collected by Mary Anning (1799-1847) held in the collections at the Natural History Museum, London. *In* Lord, A.R. & Davis, P.G. (eds), 2010. *Fossils from the Lower Lias of the Dorset coast*. Palaeontological Association Field Guide to Fossils: Number 13. London: The Palaeontological Association, 401-407. Hawkins' acknowledgement to Anning is from Hawkins, T., 1834. *Memoirs of Ichthyosauri and Plesiosauri, extinct monsters of an ancient Earth*. London: Rolfe & Fletcher, 29. It is reminiscent of Bristol fossil collector George Cumberland's remarks in 1823 and with which Hawkins was probably familiar.

38. Mantell's reservations about Hawkins' restorations are from his dairy entry of Friday 7 December 1832. See Curwen, E.C., 1940. *The journal of Gideon Mantell surgeon and geologist. Covering the years 1818-1852*. London: Oxford University Press. Mantell presents a quite contrary view of the specimen in his 1851 book *Petrifactions and their Teachings; or, a Hand-book to the Gallery of Organic Remains of The British Museum*. London: Henry G. Bohn, 381-382.

39. Anning's letter of 11 October 1833 to Charlotte Murchison is in the Geological Society Archives, LDGSL/838/A/7/3.

40. The controversy over Konig's concerns about the plaster restoration on Hawkins' specimens is summarised by Thackray, J. & Press, B., 2001. *The Natural History Museum. Nature's treasurehouse*. London: The Natural History Museum. Much more detail can be found in *Report from the Select Committee on the condition, management and affairs of the British Museum; together with the minutes of evidence, appendix and index*, 1835. London: Hansard.

41. Anna Maria Pinney (1812-1861); William Pinney (1806-1898). William Pinney was elected in 1832 and represented Lyme in Parliament until 1842, and again from 1852 to 1865. Anna Maria Pinney's journal is quoted from Lang, W.D., 1956. Mary Anning and Anna Maria Pinney. *Proceedings of the Dorset Natural History & Archaeological Society*, v.76, 146-152.

42. Lang, W.D., 1956. Mary Anning and Anna Maria Pinney. *Proceedings of the Dorset Natural History & Archaeological Society*, v.76, 147. Pinney remained in Lyme until 16 December 1831, but returned in 1832 and 1833 and spent time with Anning and the Philpot sisters.

43. Roberts, G., 1834. *The History and Antiquities of the Borough of Lyme Regis and Charmouth*. London: Samuel Bagster, 290.

44. Lang, W.D., 1956. Mary Anning and Anna Maria Pinney. *Proceedings of the Dorset Natural History & Archaeological Society*, v.76, 147.

45. Pidgeon, E., 1830. *The Fossil Remains of the Animal Kingdom*. London: Whittaker, Treacher, & Co., 377.

46. Lang, W.D., 1956. Mary Anning and Anna Maria Pinney. *Proceedings of the Dorset Natural History & Archaeological Society*, v.76, 147.

47. See note 46 above.

48. This quote is from *The Lister Thesaurus* in Lyme Regis Museum quoted by Lang, W.D., 1949. More about Mary Anning, including a newly-found letter. *Proceedings of the Dorset Natural History & Archaeological Society*, v.71, 187-188.

49. Lang, W.D., 1956. Mary Anning and Anna Maria Pinney. *Proceedings of the Dorset Natural History & Archaeological Society*, v.76, 148.

50. Charlotte Jane Skinner née Parslow (1785-1872); Maria Sarah Wallace (later Ogle) (1791-1844). Fowles, J. & Constable, J. (eds), 2011. *The Lymiad. A Poem in the form of Letters from Lyme to a Friend at Bath written during the autumn of 1818.* Lyme Regis: Lyme Regis Philpot Museum.

51. Major-General Henry Wyndham (1790-1860). For more on Henry De la Beche see McCartney, P.J., 1977. *Henry De la Beche: observations on an observer.* Cardiff: Friends of the National Museum of Wales; Sharpe, T. 2009. Slavery, sugar, and the Survey. *Open University Geological Society Journal.* v.29, 88-94; and Sharpe, T., 2013. *New insights into the early life of Henry Thomas De la Beche (1796-1855). In* Morris, R. (ed.), *A journal of Sir Henry De la Beche Pioneer Geologist (1796-1855). Written in his own hand.* Swansea: Royal Institution of South Wales.

52. Chevalier, T., 2009. *Remarkable creatures.* London: HarperCollins; Thomas, J., 2010. *Curiosity. a love story.* Toronto: McCleland & Stewart.

53. Anna Seward (1742-1809), known as the 'Swan of Lichfield'. Anning records that the quote is from 'Miss Seward's Letters'. It comes from a letter written by Seward in 1790 to the writer William Hayley (1745-1820) included in *The Letters of Anna Seward: Written Between the Years 1784 and 1807. Six volumes.* Edinburgh: Archibald Constable, 1811. Anning may have borrowed the books from one of her friends in Lyme. Seward, who never married, was herself possessed of an independent fortune. For more on Seward see Barnard, T., 2016. *Anna Seward: A Constructed Life. A Critical Biography.* Abingdon: Routledge.

54. James Montgomery (1771-1854), *The World before the Flood, a poem in ten cantos; with other occasional pieces.* London: Longman, Hurst, Rees, Orme, and Brown, 259. The quotes are contained within Anning's 'Fourth Notebook' in Dorset County Museum.

55. Sherborn C.D., 1935. Sowerby's *Mineral Conchology*: annotated list of the collectors mentioned by the Sowerbys. *Proceedings of the Malacological Society of London*, v.21: 238–243; Burek, C.V., 2001. Where are the women in geology? *Geology Today*, v.17, 110-114. Kölbl-Ebert, M., 2012. Sketching rocks and landscape: drawing as a female accomplishment in the service of geology, *Earth Sciences History*, v.31, 270-286.

56. Miss Ann Congreve (1758-1823); Mrs Sarah Tylee née Salmon (1775-1852); Mrs Elizabeth Cobbold née Knipe (1767-1824); Barbara Marchioness of Hastings (1810-1858); Lady Eliza Maria Gordon Cumming née Campbell

(c.1798-1842); Mary, 3rd Countess of Rosse (1813-1885). For more on some of these and other female collectors see Creese, M.R.S. & Creese, T.M., 1994. British women who contributed to research in the geological sciences in the nineteenth century. *British Journal for the History of Science*, v.27, 23-54; Higgs, B. & Wyse Jackson, P.N., 2007. The role of women in the history of geological studies in Ireland. *In* Burek, C.V. & Higgs, B. (eds), *The role of women in the history of geology*. The Geological Society, London, Special Publications, No.281, 137-153; and Turner, S., Burek, C.V. & Moody, R.T.J., 2010. Forgotten women in an extinct saurian (man's) world. *In* Moody, R.T.J., Buffetaut, E., Naish, D. & Martill, D.M. (eds), *Dinosaurs and Other Extinct Saurians: a Historical Perspective*. The Geological Society, London, Special Publications, No.343, 111-153.

57. Etheldred Benett (1775-1845). Her great grandfather was William Wake (1657-1737), Archibishop of Canterbury from 1716 until his death. See Torrens. H.S., Benamy, E., Daeschler, E.B., Spamer, E.E. & Bogan, A.E., 2000. Etheldred Benett of Wiltshire, England, the first lady geologist – Her fossil collection in the Academy of Natural Sciences of Philadelphia, and the rediscovery of "lost" specimens of Jurassic Trigoniidae (Mollusca: Bivalvia) with their soft anatomy preserved. *Proceedings of the Academy of Natural Sciences of Philadelphia*, v.150, 59-123; Benett, E., 1831. *A Catalogue of the Organic Remains of the County of Wiltshire*. Warminster: J.L. Vardy.

58. Burek, C.V. & Kölbl-Ebert, M., 2007. The historical problems of travel for women undertaking geological fieldwork. *In* Burek, C.V. & Higgs, B. (eds), *The role of women in the history of geology*. The Geological Society, London, Special Publications, No.281, 115-122; Kölbl-Ebert, M., 2007. The role of British and German women in early 19th-century geology: a comparative assessment. *In* Burek, C.V. & Higgs, B. (eds), *The role of women in the history of geology*. The Geological Society, London, Special Publications, No.281, 155-163. Kölbl-Ebert, M., 2001. On the origin of women geologists by means of social selection: German and British comparison. *Episodes*, v.24, 182-193. Kölbl-Ebert, M., 2002. British geology in the early nineteenth century: a conglomerate with a female matrix. *Earth Sciences History*, v.21, 3-25.

59. Some accounts of Anning's life state that she was made an honorary member of the Geological Society; she was not. Mrs Maria Graham née Dundass (1785-1842), later Lady Callcott, 1824, An Account of some Effects of the late Earthquakes in Chili. Extracted from a Letter to Henry Warburton, Esq. V.P.G.S. *Transactions of the Geological Society of London*, Series 2, v.1, 413-415. For more on the admittance of women to the Geological Society see Herries Davies, G.L., 2007. *Whatever is Under the Earth. The Geological Society of London 1807 to 2007*. London: The Geological Society; and Burek, C.V., 2009. The first female Fellows and the status of women in the Geological Society of London. *In* Lewis, C.L.E. & Knell, S.J. (eds.), *The Making of the*

Geological Society of London. The Geological Society, London, Special Publications No.317, 373-407. The Geological Society has been described as 'notoriously chauvinist' (Bunbury, T., 2016. 1847. *A Chronicle of Genius, Generosity and Savagery*. Dublin: Gill Books) but it was no more so than most other organisations of the period. It was not alone in being late to admit women members; the Royal Geographical Society admitted women only from 1913, and the Linnean Society of London in 1905. However, the Zoological Society of London had had women as members since 1826 and the Botanical Society since 1837.

60. Mary Anning to Miss M. Lister 13 January 1840 published in Lang, W.D., 1949. More about Mary Anning, including a newly-found letter. *Proceedings of the Dorset Natural History & Archaeological Society*, v.71, 184-188.

61. Presumably Anning means that there will be only one preface. Her hand-copied paper is in the Library of the Natural History Museum.

62. Cumberland, G., Some Account of the Order in which the Fossil Saurians were Discovered. *Journal of the Royal Institution*, v.27, 345-349.

63. This quote, like those above from Seward and Montgomery, is from Anning's 'Fourth Notebook', often called her Commonplace Book preserved in Dorset County Museum. See Pascoe, J., 2006. *The hummingbird cabinet. A rare and curious history of romantic collectors*. Ithaca, New York: Cornell University Press; and Goodhue. T.W., *Mary Anning's Commonplace Book* which can be found on the website of Lyme Regis Museum.

8. FOSSIL FISH AND FINANCIAL CALAMITY

1. The 11 October 1833 letter describing the death of Anning's dog is in the Geological Society Archives, LDGSL/838/A/7/3. Anning's sketch of her dog is in the Natural History Museum Library.

2. The letter describing Anning being run over by a cart is from Elizabeth Philpot to Mary Buckland, 9 December 1833, Oxford University Museum of Natural History, Buckland letters box 2, 7. It is interesting to note that Anning is out collecting on a Sunday, as her Dissenting father had done before her. Perhaps she was not yet a member of the Anglican church, or through necessity had to take every opportunity presented by tide and weather to search for fossils.

3. Lang, W.D., 1956. Mary Anning and Anna Maria Pinney. *Proceedings of the Dorset Natural History and Archaeological Society*, v.76, 150. Pinney records that she talked with Anning 'yesterday', so on Saturday 7 December, therefore before the accident on the bridge, if Elizabeth Philpot is correct that it happened the day before her letter which is dated 9 December, the Monday.

4. Lang, W.D., 1956. Mary Anning and Anna Maria Pinney. *Proceedings of the Dorset Natural History and Archaeological Society*, v.76, 148. The *Alexander*

was a Bombay trader hired by the East India Company. She left Bombay for London on 15 October 1814 and was wrecked in a storm on Chesil Beach in the early hours of 27 March 1815. Of about 150 people on board, only five crew survived. The body washed up at Lyme and attended by Mary Anning was that of a Mrs Jackson who had been returning from India with her three children. She was buried in St Michael's churchyard on 5 April 1815.

5. Reverend John Gleed (1785-1870). The succession of ministers at the Independent chapel is described in Densham, W. & Ogle, J., 1899. *The story of the Congregational churches of Dorset: from their foundation to the present time.* Bournemouth: Mate & Sons; and in Goodhue, T.W., 2001. The faith of a fossilist: Mary Anning. *Anglican and Episcopal History*, v.70, 80-100. Mike Taylor has written a useful account which awaits publication, of John Gleed as a pastor and fossil collector: Taylor, M.A., Reverend John Gleed (1785–1870), Independent minister and fossil collector of Lyme Regis. Gleed's ichthyosaur skull which was acquired by Thomas Hawkins went to the British Museum when it purchased Hawkins' collection.

6. Reverend Thomas Frederic Amelius Parry Hodges (1801-1880) studied Civil Law at Oxford, graduating BCL in 1829 and was ordained deacon that same year, and priest in 1830. A Fellow of New College, 1823-1831, he would have known Buckland, but there is no indication that he had any particular interest in geology. They did share a later interest in architecture, and Buckland chaired a meeting of the Oxford Society for Promoting the Study of Gothic Architecture on 2 February 1842 when Frederic Parry Hodges was admitted as a member. See Powell, C., 2018. *Lyme Regis Monographs.* FeedARead.com Publishing for more on T.F.A. Parry Hodges. He is not to be confused with the Reverend Thomas Hodges (*c.*1783-1847), who was Curate at St Andrew's Church, Charmouth 1818-1827. Joseph Anning probably joined the Anglican church at the same time as his sister and later served as churchwarden at St Michael's from 1844 to 1846. Despite side galleries having been added to the church in 1824, by 1832 the *Sherborne Mercury* was reporting that 'notwithstanding the erection of these galleries and many new pews, there is still a lack of seats to accommodate all who wish to attend.' See Draper, J., 2004. *Mary Anning's town – Lyme Regis.* Lyme Regis: Lyme Regis Museum, 44. For the hymns and psalms sung in St Michael's Church, see Parry Hodges, F., 1839. *A Selection of Psalms and Hymns, as chaunted and sung in the Parish Church of Lyme Regis, Dorset.* Lyme Regis: Daniel Dunster.

7. Rev. Henry Williams Rawlins (1783-1855); Rev. Francis John Rawlins (1827-1876), MA FSA. For this encounter with Anning see Lang, W.D., 1963. Mary Anning and a very small boy. *Proceedings of the Dorset Natural History & Archaeological Society*, v.84, 181-182; and Rawlins, C.W.H., 1962. *Family Quartette.* Privately published.

8. For more on Agassiz see Irmscher, C., 2013. *Louis Agassiz. Creator of American Science*. Boston and New York: Houghton Mifflin Harcourt.

9. Agassiz's remarks on the Philpot collection are in Agassiz, J.L.R., 1844. *Recherches sur les Poissons fossile*s. Neuchatel, vol.1, 22. His dedication to Elizabeth Philpot is in volume 2, part 2, 102. The specimen from Lord Cole's collection which Agassiz named *Acrodus anningiae* (now *anningae*) is in the Natural History Museum registration no. NHMUK PV P.2731. Neither *Acrodus anningiae* nor *Belenostomus anningiae* are described in Agassiz's *Recherches sur les Poissons fossile*s, but do appear in lists of species in his work, the former in volume 3, pages 175 and 385, the latter in volume 2, pages 143, 161 and 297. I am grateful to Chris Duffin for this information. Note how Agassiz refers to 'Miss Philpot' but not to 'Miss Anning', seemingly treating Elizabeth Philpot with more respect, suggesting either greater familiarity with Anning or that he regarded her as less of an equal. Anning, on the other hand, in later correspondence refers to him merely as 'Agassiz'.

10. The 1829 list of geological collections is by Robert Jameson, 1829, *Edinburgh (New) Philosophical Journal*, v.7, 113-115.

11. *Eugnathus philpotiae* has since been renamed *Furo philpotae,* as the name *Eugnathus* had already been used for a bird.

12. *Cheltenham Chronicle*, Thursday 11 February 1836, 4; *Salisbury and Winchester Journal*, Monday 8 February 1836, 3; *The Ipswich Journal*, Saturday 6 February 1836, 2; *Taunton Courier, and Western Advertiser*, Wednesday 17 February 1836, 2-3. The three-year delay in this reaching the press may be due to it coming to their attention only after Anning had been awarded a government pension in 1835.

13. Letter from Elizabeth Philpot to William Buckland 26 June 1835, private collection.

14. William Vernon Harcourt (1789-1871). Buckland's letter to Harcourt about Anning's annuity is published in Morrell, J. & Thackray, A. (eds), 1984. *Gentlemen of science, Early correspondence of the British Association for the Advancement of Science*. Camden Fourth Series Volume 30. London: Royal Historical Society. *Manchester Times*, Saturday 6 February 1836, 4. See also Goodhue, T.W., 2004. *Fossil hunter. The life and times of Mary Anning (1799-1847)*. Bethesda, Maryland: Academic Press; and Lang, W.D., 1960. Portraits of Mary Anning and other items. *Proceedings of the Dorset Natural History & Archaeological Society*, v.81, 89-91.

15. Dr John Woodward (1667-1728).

16. Reverend Thomas Hodges (c.1783-1847) of Charmouth. Two years ahead of Adam Sedgwick at Trinity College Cambridge, he graduated BA in 1806, and MA in 1809. He seems to have collected fossils and presented an ichthyosaur to Dorset County Museum in June 1846. Letter Mary Anning to Adam Sedgwick 29

NOTES, SOURCES AND FURTHER READING

June 1835; see Price, D., 1986. Mary Anning specimens in the Sedgwick Museum, Cambridge. *The Geological Curator*, v.4, 321.

17. Elizabeth Philpot told William Buckland in a letter of 26 June 1835 (private collection) that 'Mary Anning has lately found two slabs of Pentacrinite in the Oolitic sandstone near Bridport the column and heads much smaller and the species appears to be different from any that I have seen' and sent him a sketch of the specimen. Anning was correct in her identification that it was a new discovery, later named *Balanocrinus*. Egerton's observations on the ichthyosaur neck can be found in the *Proceedings of the Geological Society of London*, v.2, no.41, 192-193 and in Egerton, P. G., 1837. On certain peculiarities in the Cervical Vertebrae of the Ichthyosaurs, hitherto unnoticed. Read June 8, 1835. *Transactions of the Geological Society of London*, Series 2, v.5, 187-194.

18. Mary Anning to Adam Sedgwick 27 July 1835; See Price, D., 1986. Mary Anning specimens in the Sedgwick Museum, Cambridge. *The Geological Curator*, v.4, 321.

19. Mary Anning to Adam Sedgwick 9 September 1835; See Price, D., 1986. Mary Anning specimens in the Sedgwick Museum, Cambridge. *The Geological Curator*, v.4, 321.

20. Mary Anning to Adam Sedgwick 23 September 1835; Sedgwick spent another £7 in October 1836 buying fossils from Mary Anning, but there seems to be no record of exactly what these were. See Price, D., 1986. Mary Anning specimens in the Sedgwick Museum, Cambridge. *The Geological Curator*, v.4, 321-323.

21. The discovery of the large ichthyosaur was reported in *The Literary Gazette, and Journal of the Belles Lettres, Arts, Sciences, &c*. No. 952, Saturday, April 11, 1835. London: James Moyes, 252. The story was also carried in the *North Devon Journal*, Thursday 14 May 1835, 3; *Leicester Chronicle*, Saturday 25 April 1835, 2; *Liverpool Mercury*, Friday 1 May 1835, 2-3; *Hampshire Telegraph*, Monday 11 May 1835, 3; *Belfast News-Letter*, Friday 29 May 1835, 4; and other papers.

22. Ludwig Leichhardt (1813-c.1848). Leichhardt's visit to Lyme Regis is recorded in his letters published in Aurousseau, M. (ed.), 1968. *The letters of F.W. Ludwig Leichhardt*. 3 volumes. Cambridge: Cambridge University Press for the Hakluyt Society.

23. John Kenyon (1784-1856). *To Mary Anning* was published in Kenyon, J., 1838. *Poems: for the most part occasional*. London: Edward Moxon, 109-111. Kenyon's visit to Lyme is mentioned in Crosse, Mrs A., 1890. John Kenyon and his friends. *Temple Bar, A London Magazine for Town and Country Readers*, v.88, April 1890, 477-496. There seems to be no other evidence for the suggestion that Anning was nearly drowned in a 'water-vat'. Goodhue, T.W., 2004. *Fossil hunter. The life and times of Mary Anning (1799-1847)*. Bethesda, Maryland: Academic Press., 98.

24. Edward Charlesworth (1813-1893). Anning's letter extract is in *The Magazine of Natural History*, v.3, New Series, 1839, 605.

25. Edmund Thomas Higgins (1817-1891). For more on Higgins see Torrens, H.S., 1987. Geological collections and collectors of note. 199. Edmund Thomas Higgins (fl.1831-fl.1887). *The Geological Curator*, v.5, 30-34. The other description of its discovery is from *The Mining Journal & Commercial Gazette*, v.9 (209), 3 August 1839.

26. Charlesworth's paper describing *Hybodus delabechei* is Charlesworth, E., 1839. On the Fossil Remains of a Species of *Hybodus*, from Lyme Regis. *The Magazine of Natural History*, v.3, New Series, 242-248. Henry Woods (c.1796-1840). Anning's July 1839 letter to Charlesworth is quoted in Goodhue, T.W., 2004. *Fossil hunter. The life and times of Mary Anning (1799-1847)*. Bethesda, Maryland: Academic Press, 100.

27. I am grateful to Emma Bernard of the Natural History Museum and Simon Harris of the British Geological Survey for their help in locating Higgins' *Hybodus* specimen. At the time of writing, this specimen is on loan to Lyme Regis Museum. Higgins' donations of Lyme Regis specimens to the Yorkshire Museum include a fine, eight feet (2.4 m) long ichthyosaur 'discovered ... by the Donor'. *Annual Report of the Council of the Yorkshire Philosophical Society for 1848*, 10, 22. It is more likely, though, that he purchased it from Anning. This specimen is now in the National Museums of Northern Ireland in Belfast, having been acquired through exchange with the Yorkshire Museum in the 1970s. In 1849, through exchange, Higgins gave the Yorkshire Museum 'Fossil Fishes from Lyme Regis'. *Annual Report of the Council of the Yorkshire Philosophical Society for 1849*, 20 (as well as donating a 'Turnip presenting a singular monstrosity of form' to the botany collections).

28. Owen's letter of 13 September 1839 to Clift and his meeting with Buckland and Conybeare is quoted in Owen, R., 1894. *The life of Richard Owen*. vol. 1. London: John Murray, 164-166.

29. Anning's description of her visit to the landslip is in a letter of 13 January 1840 to Miss M. Lister and is published in Lang, W.D., 1949. More about Mary Anning, including a newly-found letter. *Proceedings of the Dorset Natural History & Archaeological Society*, v.71, 184-188.

30. William Dawson (1790-1877). Conybeare's initial account of the landslip is dated 31 December 1839 and was published in *Wolmer's Exeter and Plymouth Gazette* and in the *Taunton Courier* and was reprinted separately by Dunster on Broad Street in Lyme Regis. It can found also as Conybeare, W.D., 1840. Extraordinary land-slip and great convulsion of the coast of Culverhole Point, near Axmouth. *The Edinburgh New Philosophical Journal*, v.29, July 1840, 160-164. See also Conybeare, W.D., Dawson, W., Buckland, W., Buckland, M., 1840. *Ten plates comprising a plan, sections, and views, representing the changes*

produced on the coast of East Devon, between Axmouth and Lyme Regis by the subsidence of the land and elevation of the bottom of the sea, on the 26th December, 1839, and 3rd of February, 1840. London: John Murray; and Roberts, G., 1840. *An account of and guide to the Mighty Landslip of Dowlands and Bindon in the parish of Axmouth near Lyme Regis, December* 25, 1839. Lyme Regis: Daniel Dunster. For a modern account see Pitts, J. & Brunsden, D., 1987. A reconsideration of the Bindon Landslide of 1839. *Proceedings of the Geologists' Association,* v.98, 1-18; and for an appreciation of the work of Conybeare and Buckland see Gallois, R., 2010. The failure mechanism of the 1839 Bindon Landslide, Devon, UK: almost right first time. *Geoscience in South-West England,* v.12, 188-197.

9. 'I AM WELL KNOWN THROUGHOUT THE WHOLE OF EUROPE'

1. Mantell's comments on the 25 July 1844 auction are in Curwen, E.C., 1940. *The journal of Gideon Mantell surgeon and geologist. Covering the years* 1818-1852. London: Oxford University Press.

2. I am grateful to Hellen Pethers of the Natural History Museum Library for access to the museum's catalogues. The starfish purchased from Anning are the brittle-stars *Palaeocoma egertoni* and *Ophioderma tenuibrachiata,* museum registration numbers NHMUK 14398-40; the fish teeth belong to *Hybodus* cf. *delabechei,* a species described by Edward Charlesworth in 1839 and named after Henry De la Beche, museum number NHMUK 20600a. Details can be found in Chapman, S. & Milner, A.C., 2010. Appendix. A catalogue of fossil specimens collected by Mary Anning (1799-1847) held in the collections at the Natural History Museum, London. *In* Lord, A.R. & Davis, P.G. (eds), 2010. *Fossils from the Lower Lias of the Dorset coast.* Palaeontological Association Field Guide to Fossils: Number 13. London: The Palaeontological Association, 401-407.

3. Hugh Edwin Strickland (1811-1853). The quote about Anning and the possible ammonite ink-sacs is from Strickland, H.E., 1845. On certain calcareocorneous bodies found in the outer chamber of ammonites. *Quarterly Journal of the Geological Society,* v.1, 232-235.

4. Sadly, Strickland came to an unfortunate end in 1853 at the age of 42. While examining rocks exposed in a curved railway cutting in Nottinghamshire, he moved out of the way of an approaching coal train and into the path of a passenger train on the other line. He was killed instantly. Jardine, W., 1858. *Memoirs of Hugh Edwin Strickland, M.A.* London: John van Voorst, cclix.

5. *Taunton Courier, and Western Advertiser,* Wednesday 19 October 1842, 7.

6. Eleanor Emma Waring (1838-1909), *Peeps into an old playground. Memoirs of the past.* 1895. Quoted by Lang, W.D., 1949. More about Mary Anning, including a newly-found letter. *Proceedings of the Dorset Natural History &*

Archaeological Society, v.71, 187.

7. Anning's commonplace book, which now belongs to Dorset County Museum in Dorchester, was dismissed as typical of a young lady of the period and of no value by Harold Idris Bell (1879-1967), Keeper of Manuscripts at the British Museum in declining the offer of the book in 1935 by Charles Davies Sherborn (1861-1942) of the British Museum (Natural History) Department of Geology. W.D. Lang considered that it pointed to her simplicity and piety in Lang, W.D., 1949. More about Mary Anning, including a newly-found letter. *Proceedings of the Dorset Natural History & Archaeological Society*, v.71, 188. For recent reassessments of the book and an explanation of its contents see Pascoe, J., 2006. *The hummingbird cabinet. A rare and curious history of romantic collectors.* Ithaca, New York: Cornell University Press; and Goodhue. T.W., *Mary Anning's Commonplace Book* which can be found on the website of Lyme Regis Museum.

8. George Gordon Byron (1788-1824), Henry Kirke White (1785-1806), Tobias Smollett (1721-1771), Thomas Gray (1716-1771) and Hannah More (1745-1833); Thomas Wilson, Bishop of Sodor and Man (1663-1755).

9. Mary Anning to Adam Sedgwick 4 May, 20 May, 26 May 1843; see Price, D., 1986. Mary Anning specimens in the Sedgwick Museum, Cambridge. *The Geological Curator*, v.4, 322-323; Reverend Osmond Fisher (1817-1914).

10. Dorothea Solly (c.1787-1878); Reverend Thomas Rackett (1787-1840); Samuel Solly (c.1781-1847) joined the Geological Society in 1810 and served on its Council 1811-1819; he was elected FRS in 1812. Not to be confused with the merchant Samuel Solly FRS (c.1724-1807) or the surgeon Samuel Solly FRS (1805-1871). Anning's letter to Dorothea Solly of 10 June 1844 was published in Lang, W.D., 1953. Mary Anning and the fire at Lyme in 1844. *Proceedings of the Dorset Natural History & Archaeological Society*, v.74, 175-177. Although Lang cites Mary Anning's notes on the differences between living and fossil species as her own essay, it is in fact copied verbatim from a footnote in the introduction to Conybeare, W.D. & Phillips, W., 1822. *Outlines of the geology of England and Wales*. London: William Phillips, ix, x.

11. The *Illustrated London News* report and picture of the fire is in the 18 May 1844 issue. The quote of Lyme ascending in flames to heaven is from Wanklyn, C., 1927. *Lyme Regis A retrospect*. Second edition. London: Hatchards, 103.

12. Frederick Augustus II (1797-1854), King of Saxony; Carl Gustav Carus (1789-1869). The visit of the King of Saxony to Lyme is described in Carus, C.G., 1846. *The King of Saxony's journey through England and Scotland in the year 1844. Translated by S.C. Davidson*. London: Chapman & Hall. For a railway-centred account of their journey see Barnes, R., 2002. A Royal Progress. *BackTrack. Recording Britain's Railway History*, v.16, 346-351, 406-413.

13. It is unclear from Carus' remarks on the reasonably-priced ichthyosaur whether he or the king bought the specimen, but his expression of being 'anxious,

at all events, to write down the address' suggests that the specimen was not purchased. Anning wrote her name in Carus' notebook, not the king's as is often misreported. Carus, in his published account of the visit, records the name as 'Annins' but this is a misreading of 'g' as 's'.

14. Leopold von Buch (1774-1853). A transcript and translation of von Buch's lecture and remarks about Mary Anning can be found in Kröger, B., 2013. Remarks on a scene, depicting the primeval world: a talk given by Leopold von Buch in 1831, popularizing the Duria antiquior. *HiN – Humboldt Im Netz. Internationale Zeitschrift für Humboldt-Studien*, XIV, v.27, 7-35.

15. Samuel Griswold Goodrich (1793-1860), *Peter Parley's Wonders of the Earth, Sea and Sky*. New York: S. Colman. 1840, 17. The English edition was edited by the Rev. T. Wilson and was published in London by Darton and Clark.

16. Thomas Wilson (1663-1755). Anning may have copied Wilson's prayers from Cruttwell, Rev. C., 1782. *The works of the Right Reverend Father in God Thomas Wilson, D.D. fifty-eight years Lord Bishop of Sodor and Man. With his life compiled from authentic papers*. 2nd edition. Vol 1. Bath. For more on her commonplace book notes and prayers see Goodhue, T.W., 2001. The faith of a fossilist: Mary Anning. *Anglican and Episcopal History*, v.70, 80-100; and Goodhue, T.W., 2004. *Fossil hunter. The life and times of Mary Anning (1799-1847)*. Bethesda, Maryland: Academic Press. Mention of Anning taking opium is in Anon. [H.S. Fagan], 1865. 'Mary Anning, the fossil finder'. *All the year round*, v.13 (303) (11 February), 62, a source which has been shown to be unreliable.

17. S.E. [Eleanor Emma Waring (1838-1909)], 1895. *Peeps into an old playground. Memories of the past*. Lyme Regis: F. Dunster, 4. Also quoted by Lang, W.D., 1949. More about Mary Anning, including a newly-found letter. *Proceedings of the Dorset Natural History & Archaeological Society*, v.71, 184-188.

18. For Buckland's organising of a second subscription for Anning see Buckland's letter of 10 October [1846] to Sir Walter Calverley Trevelyan (1797-1879), British Library, Add MSS 31026/247. The report on the subscription launched for Anning is reported in *The Manchester Times*, Friday 30 October 1846, 5.

19. The minute books of the Dorset County Museum 2 July 1846 'Resolved that Miss Mary Anning and ... be requested to become Honorary Members of the Institution'. See Lang, W.D., 1939. Mary Anning (1799-1847) and the pioneer geologists of Lyme. *Proceedings of the Dorset Natural History & Archaeological Society*, v.60, 162. Some modern accounts of Anning's life erroneously report this as honorary membership of the Geological Society of London, for example Uglow, J. (ed.), 1998. *The Macmillan Dictionary of Women's Biography*. Third Edition. London: Macmillan, 22. The Rev. Thomas Hodges presented Dorset County Museum with its first geological specimen, an *Ichthyosaurus communis*

(accession number G.1) may have been collected originally by Mary Anning, or perhaps by one of the Charmouth fossilists.

20. The printers of Anning's stationery, which was probably available from provincial stationers of which there were several in Lyme, operated under the name of J. & F. Harwood between 1830 and 1843. The two poems were first published in *Bentley's Miscellany*, v.19, 312-314. This was a literary magazine started by Richard Bentley (1794-1871) and published from 1836 until 1868. The two poems, were republished in 1869 in Sheehan, J., 1869. *The Bentley Ballads, comprising the Tipperary Hall Ballads, now first republished from "Bentley's Miscellany"*. Third edition. London: Richard Bentley, the preface to which identifies the author as a young Cambridge University physician.

21. Like his sister, Joseph became a prominent figure in Lyme, with his business as an upholsterer and cabinet-maker, and played an active role in public life in the town. In 1835 he was Overseer of the Poor, in charge of dispensing the aid from which he and his sister and mother had benefited just twenty years before. The responsibilities of this post forced him to neglect his business, especially as his assistant had been dismissed the year before because of irregularities in the accounts (Letter from Joseph Anning to John George Shaw Lefevere, Poor Law Commissioner 17 September 1835 in the National Archives, Kew, MH 12/2095/14 Folio 24). His children buried here are probably William Henry Anning (1833-1838) and Joseph Reader Anning (1836-38); the third child is unrecorded but may be a stillborn baby or one who died before being baptised. Although we know that Mary Anning was buried at St Michael's, there is no definitive record that the named stone marks her grave. It is possible that she may have been buried elsewhere in the churchyard in a plot now unknown, and is commemorated by this stone. Lang, W.D., 1949. More about Mary Anning, including a newly-found letter. *Proceedings of the Dorset Natural History & Archaeological Society*, v.71, 188 notes that it was suggested in about 1919 that the gravestone lettering be renewed.

22. Notice of Anning's death appeared in *The Athenaeum Journal of Literature, Science, and the Fine Arts, for the year 1847*, No. 1013, Saturday 27 March 1847, 337; as well as in *The Gentleman's Magazine*, v.27, New Series, May 1847, 562; the *Hereford Journal*, Wednesday 31 March 1847, p.3, col.3; *The Essex Standard, and Eastern Counties Advertiser*, Colchester, Essex, 9 April 1847; *Glasgow Herald*, Monday 5 April 1847, 4; *The Pilot*, Monday 5 April 1847, 4; *Carlisle Patriot*, Friday 2 April 1847, 3; *Bury and Norwich Post*, Wednesday 7 April 1847, 1; *Salisbury and Winchester Journal*, Saturday 3 April 1847, 3; *London Daily News*, Tuesday 30 March 1847, 5; *Northern Star and Leeds General Advertiser*, Saturday 3 April 1847, 9; *Ipswich Journal*, Saturday 3 April 1847, 3; and other newspapers, including *Home News for India, China and the Colonies*, Wednesday 7 April 1847, 13. George Roberts left an unpublished 1847

manuscript, *A brief memoir of Miss Mary Anning, the celebrated fossilist* now preserved within the Eyles MSS, Bristol University Library.

23. De la Beche's obituary of Mary Anning was published in the *Quarterly Journal of the Geological Society*, v.4, xxi-cxx, 1848. It was in part based upon a letter he received, at the instigation of George Roberts, from Anning's thirteen-year old nephew, Charles Churchill Anning (De la Beche Archive, National Museum of Wales, NMW 84.20G.D11).

24. Augustus W.N. Pugin (1812-1852); William Wailes (1808-1881). The main elements of the St Michael's window's design, the six quatrefoils, are almost identical to those in a window in St Giles' Roman Catholic Church in Cheadle in Staffordshire which was built between 1841 and 1846. For more on the Acts of Mercy depicted in stained glass windows see Brocket, J., 2018. *How to look at stained glass. A guide to the church windows of England*. London: I.B. Tauris & Co.

25. The stained glass window to Mary Anning is the westernmost window in the north aisle, opposite the Anning gravestone. George Roberts' annotated copy of his 1834 *History and Antiquities of the Borough of Lyme Regis and Charmouth*, gives the date of the window's installation as February 1850 and mentions a subscription for a tablet. See Lang, W.D., 1960. Portraits of Mary Anning and other items. *Proceedings of the Dorset Natural History & Archaeological Society*, v.81, 89-91. I am grateful to Caroline Lam, Geological Society Archivist for checking the Society's minutes for a mention of the window subscription. The Society did, however, contribute to recent repairs to the window.

26. At the time of his death Conybeare was Dean of Llandaff Cathedral in Wales. The deaths were reported in the *Taunton Courier and Western Advertiser* on Wednesday 9 September 1857, 7.

10. LEGACY AND LEGEND

1. Henry Rowland Brown (1837-1921). Brown's remarks about Anning are in Brown, H.R., 1859. *The beauties of Lyme Regis, Charmouth, the Land-slip, and their vicinities*. 2nd edition. London: Longman. This edition contains an expanded piece about the life of Mary Anning over that in the 1857 first edition. A recent reassessment of Brown's writing on Anning and its later plagiarism can be found in Taylor, M.A. & Torrens, H.S., 2014. An account of Mary Anning (1799-1847), fossil collector of Lyme Regis, Dorset, England, published by Henry Rowland Brown (1837-1921) in the second edition (1859) of *Beauties of Lyme Regis*. *Proceedings of the Dorset Natural History and Archaeological Society*, v.135, 62-70.

2. Brown, H.R., 1857. *The beauties of Lyme Regis, Charmouth, the Land-slip, and their vicinities*. Lyme Regis: Daniel Dunster, 28, footnote and Advertisements, 28. Henry Marder (born c.1818); James Marder (1824-1888). Anna Maria Pinney

notes in her journal that on Wednesday 21 November 1832 while on a fossiling excursion with Mary Anning to the east of Charmouth 'little Marder went with us'. This was most likely Henry Marder, then about 14 years old. Anning was training a new generation of fossilist.

3. For more on fossil collecting in Yorkshire and the Brown Marshalls, father and son, see Knell, S.J., 2000. *The culture of English Geology, 1815-1851. A science revealed through its collecting.* Aldershot: Ashgate, 122-123, 198-210.

4. William Moore (c.1795-1867); Sarah Moore (c.1795-1865); Moore's letter to De la Beche is in the Department of Geology, National Museum of Wales, NMW84.20G.D988. De la Beche may not have purchased the ichthyosaur; no specimen matching its description seems to be present in the collections of the British Geological Survey.

5. After his wife's death, Moore married Harriet Symes (*baptised* 1829-1892) in 1866, but he himself died the following year. In 1871 Harriet married James Dollin (1811-1896) who had been a fisherman and innkeeper of the Ship Inn in Horse Street (now the northern part of Coombe Street) and continued the fossil business until the 1890s after which it passed to Thomas Seager who is listed as a fossil dealer in an 1895 directory. Charles Rennie Mackintosh (1868-1928). Mackintosh's pencil drawing of the Fossil Depot is now in the collections of the Hunterian Art Gallery of Glasgow University. The suggestion that the whale bone was displayed outside Anning's shop is in Draper, J., 2004. *Mary Anning's town – Lyme Regis.* Lyme Regis: Lyme Regis Museum, 24. Sidney Curtis (*died* 1926); Sir Frederick Treves (1853-1923).

6. Charmouth fossil collectors included John Hunter (1794-1877); Isaac Hunter (1833-1906); and Samuel Clark (1815-1888). The report of the Geologists' Association visit to Isaac Hunter is in Woodward, H.B., 1906. Excursion to Lyme Regis, Easter, 1906. *Proceedings of the Geologists' Association*, v.19, 320-340.

7. Proby Thomas Cautley (1802-1871). Mary Anning's copy of Roberts' *The History of Lyme-Regis,* signed by her and dated 1823, is in Cardiff University Library where it was located in 2005 by Chris Powell to whom I am grateful for this information. Powell noted that there were three pressed flowers in the book which might just have been put there by Mary Anning herself.

8. John Phillips (1800-1874), palaeontologist, Professor of Geology at King's College London from 1834, at Dublin from 1844, and was also attached to the Geological Survey founded in 1835 by Henry De la Beche. Philips, J., 1865-70. *A Monograph of British Belemnitidae.* London: The Palaeontographical Society, 40, 42, 50, 51, 53 cite specimens from Anning. More on Phillips' work on belemnites can be found in: Morrell, J., 2005. *John Phillips and the business of Victorian science.* Aldershot: Ashgate Publishing. Three letters written in 1835 from Mary Anning to John Phillips about his enquiry and purchase of belemnite fossils are in the Phillips Letters, Oxford University Museum of Natural History. I am grateful

to Eliza Howlett at the OUMNH for bringing these to my attention.

9. Charles Davies Sherborn (1861-1942). See Norman, J.R., 1944. *Squire. Memories of Charles Davies Sherborn*. London: George G. Harrap & Co., 88. Other Anning correspondence occasionally resurfaces, such as the previously unknown letter from Anning to William Buckland dated 15 February [1830] put up for sale at Sotheby's in London in August 2020. This probably came from Buckland's descendants who released some papers in the 1930s, rather than from the papers distributed by Sherborn.

10. Richard Owen MSS 149A, 149B, Natural History Museum Library Enniskillen to Richard Owen, 25 August 1885; Albert Anning to Lord Enniskillen 13 August 1885.

11. Taylor, M.A., 1986. The Lyme Regis (Philpot) Museum: the history, problems and prospects of a small museum and its geological collection. *The Geological Curator*, v.4, 313.

12. For more on these hammers see Palmer, D., 2016. *Tools of the trade. The Sedgwick Museum's historical collection of geological hammers*. Cambridge: Sedgwick Museum.

13. Cyril Wanklyn (1864-1943); John Fowles (1926-2005); Jo Draper (1949-2017).

14. T.E.D. Philpot (1859-1918) was Mayor of Lyme Regis 1891-93. After his death in 1918, the museum was given to the town in 1920 by his niece. For more on T.E.D. Philpot and the building of Lyme Regis Museum see Hebditch, M., 2018. *The architect George Vialls and his clients in late 18th century Lyme Regis* which can be found on the Lyme Regis Museum website. See also, Taylor, M.A., 1986. The Lyme Regis (Philpot) Museum: the history, problems and prospects of a small museum and its geological collection. *The Geological Curator*, v.4, 309-317.

15. For a study of the Anning medallion or token, see Taylor, M.A., & Bull, R., 2015. A token found at Lyme Regis, Dorset, England, apparently associated with Mary Anning (1799-1847), fossil collector. *Proceedings of the Dorset Natural History and Archaeological Society*, v.136, 62-67.

16. The investigations of Richard Bull into Mary Anning's fossil extractor are described in Bull, R., 2013. Lyme's history in objects. 15. 1888-92 Army entrenching tool handle – once thought to be 'Mary Anning's Hammer' which can be found on the Lyme Regis Museum website.

17. William Gray (1815-1885).

18. The portrait passed to Amelia's youngest son, Albert (b.1842) and then to the sixth of Albert's nine children, Annette (1876-1938). She presented it to the British Museum (Natural History), seemingly after it had been offered for sale to, and turned down by, the National Portrait Gallery, the British Museum (Natural History) and the Lyme local historian Cyril Wanklyn. For more on

the Anning portraits see Lang, W.D., 1960. Portraits of Mary Anning and other items. *Proceedings of the Dorset Natural History & Archaeological Society*, v.81, 89-91.

19. Anning's sketch of her dog is amongst her papers on the Owen archive in the Natural History Museum Library. The name of Anning's dog, Tray, comes from two letters by Annette Anning, Mary Anning's great-niece, quoted in Lang, W.D., 1949. More about Mary Anning, including a newly-found letter. *Proceedings of the Dorset Natural History & Archaeological Society*, v.71, 184-188.

20. Benjamin John Merefield Donne (1831-1928). Wanklyn, C., 1927. *Lyme Regis A retrospect*. Second edition. London: Hatchards, xviii.

21. We have no record of who owned the picture or where it was between 1850 and 1875. In a letter dated 9 June 1875 to the Secretary of the Geological Society, James Marder, Broad St, Lyme Regis, asks for a refund of his expenditure in sending the portrait of Mary Anning, a gift from Lord Enniskillen, to the Geological Society. Geological Society Archives GSL/L/R/19/151. On 21 October 1922, Lyme Historian Cyril Wanklyn wrote to Leo Baptiste Louis Belinfante, the Geological Society's Permanent Secretary to inform him that he had located B.J.M. Donne still living at Merifield, Louisa Terrace, Exmouth. I am grateful to Caroline Lam, Geological Society Archivist for her help with this.

22. Thomas Wyse (1791-1862). Lang suggests that 'Miss Penning' may be Anna Maria Pinney, but Roberts' has clearly written 'Penning' in his notes. Amongst the subscribers to his 1834 book was a Mrs Penning who may be connected with the lady who wanted an Anning portrait.

23. Thomas Carter Galpin (1795-1850).

24. This drawing seems to have been first identified as Mary Anning in North, F.J., 1933. Dean Conybeare, Geologist. *Transactions of the Cardiff Naturalists' Society*, v.66, plate 2A; its attribution to De la Beche dates from the 1980s. See Sharpe, T., 2021. A case of mistaken identity: is Mary Anning (1799-1847) actually William Buckland (1784-1856)? *Earth Science History*, v.40, 1-16.

25. Draper, J., 2006. *Lyme's first photographers*. Lyme Regis: Lyme Regis Museum.

26. William Henry Fox Talbot (1800-1877). The 1843 photograph purporting to be of Anning and De la Beche is discussed thoroughly (and dismissed) by Taylor, M.A. & Levitt, S., 2015. Mary Anning (1799-1847) and the photograph *The Geologists* ascribed to William Henry Fox Talbot (1800-1877). *Geoscience in South-West England*, v.13, 419-427.

27. Maria Hack (1777-1844). 1832. *Geological Sketches, and Glimpses of the Ancient Earth*. London: Harvey and Darton, 294-304. The 'forgotten' Anning quotes are from Monroe, J.S. & Wicander, R., 2009, *The changing Earth. Exploring geology and evolution*, fifth edition, Belmont, California: Brooks/Cole, 622; and Bunbury, T., 2016. 1847 *A Chronicle of Genius, Generosity and*

NOTES, SOURCES AND FURTHER READING

Savagery. Dublin: Gill Books. She was, however, remembered by Mantell, G.A., 1851. *Petrifactions and their teachings, or, A hand-book to the gallery of organic remains of the British Museum.* London: Henry G. Bohn, 367; and by Sedgwick, A., 1869. Prefatory Notice. *In* Seeley, H.G., *Index to the Fossil Remains of Aves, Ornithosauria and Reptilia, from the Secondary System of Strata arranged in the Woodwardian Museum of the University of Cambridge.* Cambridge: Deighton, Bell, and Co., iv. Owen's tribute to Anning is on page 80 of his *Monograph of the Fossil Reptilia of the Liassic Formations. Part Second. Pterosauria.* London: The Palaeontographical Society, 1870. Robert Fisher Tomes (1823-1904); Tomes, R.F., 1878. On the stratigraphical position of the corals of the Lias of the Midland and Western counties of England and of South Wales. *Quarterly Journal of the Geological Society*, v.34, 179-195.

28. Robert Broom (1866-1951). Broom, R., 1927. On a new type of mammal-like reptile from the South African Karroo Beds (*Anningia megalops*). *Proceedings of the Zoological Society of London*, v.97, 227-232. In a subsequent publication, Broom, R., 1932. *The mammal-like reptiles of South Africa.* London: H.F. & G. Witherby, he proposed a new fossil reptile order, the Anningiamorpha. Unfortunately, Broom's new genus was based on an incomplete and poorly preserved skull now in the Transvaal Museum and *Anningia* and the Anningiamorpha are now no longer regarded as a valid names by palaeontologists. See Reisz, R.R. & Dilkes, D.W., 1992. The taxonomic position of *Anningia megalops*, a small amniote from the Permian of South Africa. *Canadian Journal of Earth Sciences*, v.29, 1605-1608.

29. Leslie Reginald Cox (1897-1965). Cox, L.R., 1936. The Gastropoda and Lamellibranchia of the Green Ammonite Beds of Dorset. *Quarterly Journal of the Geological Society of London*, v.92, 456-471; Cox, L.R., 1958. Anningella, nom.nov. for *Anningia* Cox *non* Broom. *Proceedings of the Geological Society of London*, No.155, 44.

30. *Cytherelloidea anningi* was described in Lord, A., 1974. Ostracods from the Domerian and Toarcian of England. Palaeontology, v.17, 599-622; the plesiosaur is described in Vincent, P. & Benson, R.B.J., 2012. *Anningasaura*, a basal plesiosaurian (Reptilia, Plesiosauria) from the Lower Jurassic of Lyme Regis, United Kingdom. *Journal of Vertebrate Paleontology*, v.32, 1049-1063; and the ichthyosaur in Lomax, D.R. & Massare, J.A., 2015. A new species of *Ichthyosaurus* from the Lower Jurassic of West Dorset, England, UK. *Journal of Vertebrate Paleontology*, v.35, 1-14.

31. W[oodward], B.B., 1909. Anning, Mary (1799-1847). *Dictionary of National Biography*, XXII (Supplement). London: Smith, Elder & Co., 51-52. Bernard Barham Woodward (1853-1930). Anning's *DNB* entry was updated by Hugh Torrens in 2004. An entry under 'Anning, Mary' can be found in *The Encyclopaedia Britannica*, 11th edition, 1910, v.2, 74.

32. Newspaper articles featuring Anning include *Bridport News*, Friday 14 May 1880, 4; *Express and Echo*, Wednesday 17 September 1884, 1; *Bournemouth Daily Echo*, Friday 4 August 1905, 4; *Newcastle Daily Chronicle*, Tuesday 28 January 1908, 6; *Northern Whig*, Saturday 15 February 1908, 2; *Westminster Gazette*, Wednesday 28 December 1921; *Worthing Gazette*, Wednesday 29 February 1928, 5; *Littlehampton Gazette*, Friday 2 March 1928, 5; *Northern Whig*, Wednesday 10 July 1929, 5. Bickley, F., 1911. *Where Dorset meets Devon*. London: Constable, 55 calls Anning 'Saint Georgina of Lyme'. She is also mentioned in travel guides including Heath, Frank R., 1922. *Dorset*. 6th edition. London: Methuen & Co.; Michelin Travel Publications, 2000. *The Green Guide, The West of England and the Channel Islands*, 165, 224; and Richards, R., 2019. *Slow Travel. Dorset*. Third edition. Chalfont St Peter: Bradt Travel Guides, 152-153. Henry Rowland Brown, 1857 and 1859. *The beauties of Lyme Regis, Charmouth and the landslip, and their vicinities; topographically and historically considered*. Lyme Regis: Daniel Dunster, 26-30 (first edition); Longman, London, 55-64 (second edition); see Taylor, M.A. & Torrens, H.S., 2014. An account of Mary Anning (1799-1847), fossil collector of Lyme Regis, Dorset, England, published by Henry Rowland Brown (1837-1921) in the second edition (1859) of *Beauties of Lyme Regis*. Proceedings of the Dorset Natural History and Archaeological Society, v.135, 62-70. Frank Buckland (1826-1880). Anon., 1857. The fossil-finder of Lyme-Regis. *Chambers's Journal of Popular Literature, Science and Arts*, v.7, 382-384; this was reprinted in a New York magazine: Anon., 1858. The fossil-finder of Lyme Regis. *Eclectic Magazine*, v.43, 261–263; see Taylor, M.A. & Torrens, H.S., 2014. An anonymous account of Mary Anning (1799-1847), fossil collector of Lyme Regis, England, published in *Chambers's Journal* in 1857, and its attribution to Frank Buckland (1826-1880), George Roberts (c.1804-1860) and William Buckland (1784-1856). Archives of Natural History, v.41, 309-325.

33. Anon. [H.S. Fagan], 1865. 'Mary Anning, the fossil finder'. *All the year round*, v.13 (303) (11 February), 60-63. Charles Dickens (1812-1870); Elizabeth Cleghorn Gaskell (1810-1865); Anthony Trollope (1815-1882); William Wilkie Collins (1824-1889); Henry Stuart Fagan (1827-1890). For a critical assessment of this article, see Taylor, M.A. & Torrens, H.S., 2014. An anonymous account of Mary Anning (1799-1847), fossil collector of Lyme Regis, Dorset, England, published in *All The Year Round* in 1865, and its attribution to Henry Stuart Fagan (1827-1890), schoolmaster, parson and author. Proceedings of the Dorset Archaeological and Natural History Society, v.135, 71-85.

34. Versions of Fagan's article, entitled 'The female fossil finder' appear in, for example, *The Scotsman*, 3 March 1865, 3; *Kentish Chronicle*, 18 February 1865, 6; and in French, 'Mary Anning'. *Revue britannique, Revue Internationale. Choix d'articles extraits des meilleurs écrits périodiques de la Grande-Bretagne*

et de l''amérique. Tome 3, 385-395, 1865. Thorne, Isabel, 1869. The little fossil-gatherer. *Chatterbox*, no.49, 2 November 1869, 386-387. *Chatterbox* was founded in 1866 by the Reverend John Erskine Clarke (1827-1920) and publication continued until the 1950s. The cover illustration may be by a wood engraver named Johnson. The author has not yet been identified, but may be Isabel Jane Thorne née Pryer (1834-1910). The verse is in *Lyme Lyrics*, Printed and Published by E. Locke, Bridge Street, Lyme Regis, 1884 and is quoted from Lang, W.D., 1949. More about Mary Anning, including a newly-found letter. *Proceedings of the Dorset Natural History & Archaeological Society*, v.71, 186.

35. Harriet Anne Forde (1834-1896). *Mary Anning. The heroine of Lyme Regis*. London: Wells Garner, Darton & Co. 1925.

36. These children's books include H.B. Bush, 1960. *Treasures in the Rock*. Toronto: Longmans Green, republished until at least 1976 under various titles; Cole, S.R, 1991. *The dragon in the cliff: a novel based on the life of Mary Anning*. New York: Lothrop, Lee & Shepard; Day, M., 1992. *Dragon in the rocks: a story based on the childhood of the early paleontologist Mary Anning*. Toronto: Greey de Pencier Books; Goodhue, T., 2002. *Curious bones: Mary Anning and the birth of paleontology*. Greensboro, N.C.: Morgan-Reynolds Publishing; Brown, D., 2003. *Rare treasure: Mary Anning and her remarkable discoveries*. Houghton Mifflin; Brighton, C., 2007. *The fossil girl*. Lincoln Children's Books; Redmond, S.R., 2012. *The dog that dug for dinosaurs*. Simon Spotlight; Pearson, D., 2015. *Mary Anning: the girl who cracked open the world*. Pearson Scott Foresman; Claybourne, A., 2017. *Mary Anning fossil hunter*. Collins; Snedden, R., 2019. *Science superstars. Mary Anning*. Raintree; Simmons, A., 2019. *Lightning Mary*. Andersen Press. Hugh Torrens published an excellent summary of Anning's rise to popular fame in his much-cited 1995 address to the British Society for the History of Science: Torrens, H.S., 1995. Mary Anning (1799-1847) of Lyme; 'the greatest fossilist the world ever knew'. *British Journal for the History of Science*, v.28, 257-284.

37. Lang's comment is from his 1949 paper, More about Mary Anning, including a newly-found letter. *Proceedings of the Dorset Natural History and Archaeological Society*, v.71, 188.

38. Anning inspires the fossiling interest of the lead character in Perry, S., 2016. *The Essex Serpent*. London: Serpent's Tail. Five biographical accounts of Anning's life have also been published during the last 25 years: Tickell, C., [1996]. *Mary Anning of Lyme Regis*. Lyme Regis: Lyme Regis Philpot Museum; Lowton, R.M., 1997. *Mary Anning. Family and friends*. Lyme Regis; Goodhue, T.W., 2004. *Fossil hunter. The life and times of Mary Anning (1799-1847)*. Bethesda, Maryland: Academic Press; Pierce, P. 2006. *Jurassic Mary. Mary Anning and the primeval monsters*. Stroud: Sutton Publishing; Emling, S., 2009. *The fossil hunter. Dinosaurs, evolution, and the woman whose discoveries changed the world*. New

York: Palgrave Macmillan.

39. *Fossil Woman*, written by Alison Edgar, Helena Uren and Louise Warren, and presented by Shaker Productions and Alarmist Theatre ran in London in 1996 and again in January 1998 at the Lyric Theatre Studio in Hammersmith. It was reviewed in *The Stage*, Thursday 19 December 1996, 10 and was *TimeOut* Critic's Choice for 1996. The 1998 performance was reported in *Geology Today*, v.14, 4, January-February 1998 and in the *Uxbridge & W.Drayton Gazette*, Wednesday 14 January 1998, 25. *She sells sea shells* by writer and director Helen Eastman was performed at the Clapham Omnibus Theatre in London and at the Geological Society before moving to Edinburgh; *The excavation of Mary Anning* is a 2017 play by Ian August; *Mary Anning – the mad woman of Lyme!* is by Peter John Cooper, for the AsOne Theatre Company, and is a multi-media reworking of his 2012 play, *The cabinet maker's daughter*, for the 2020 Lyme Regis Fossil Festival.

40. *The Crocodile* was a three-part play by John Tully for BBC Schools programme *Merry-go-Round* and first broadcast in 1973; *The French Lieutenant's Woman*, directed by Karel Reisz and adapted by Harold Pinter starred Jeremy Irons and Meryl Streep; *Mary Anning and the dinosaur hunters* is a two-part independent film production by Mermade Films, written and directed by Sharon Sheehan; *Ammonite* is a See-Saw Films production written and directed by Francis Lee and stars Kate Winslet as Anning.

41. Harry Gifford (1876-1960); Wilkie Bard (1870-1944).

42. See Winick, S., 2017. She sells seashells and Mary Anning: metafolklore with a twist. *Folklife Today*. https://blogs.loc.gov/folklife/2017/07/she-sells-seashells-and-mary-anning-metafolklore-with-a-twist/#locshare/print. The earliest publication of the four word line, *She sells sea shells*, Winick found is in Bell. A.M., 1855. *Letters and sounds: an introduction to English reading, on an entirely new plan*. London: Hamilton, Adams & Co. Alexander Melville Bell (1819-1905) was the father of telephone inventor Alexander Graham Bell (1847-1922). For more on Wilkie Bard see https://www.mislaidcomedyheroes.com/wilkie-bard. In her 1894 biography of her father, William Buckland's daughter, Elizabeth (1837-1919) implies that Anning sold 'pretty little boxes of shells or tastefully arranged bunches of seaweed' but likely confuses Anning's shop with the later Fossil Depot at Cobb Gate. See Gordon, Mrs [Elizabeth Oke Buckland], 1894. *The life and correspondence of William Buckland, D.D., F.R.S.* London: John Murray, 116. For its educational usage see Clary, R.M. & Wandersee, J.H., 2006. Mary Anning: she's more than "seller of sea shells at the seashore". *The American Biology Teacher*, v.68, 153-157.

43. Taylor, M.A., 1999. The greatest fossilist the world ever knew: Mary Anning (1799-1847). *Endeavour*, v.23, 93-94. For a report on the 1999 conference see Howe, S.R., 1999. Mary Anning and her times: a bicentenary meeting, Lyme Regis,

Dorset 2-4 June 1999. *Coprolite*, no.30, 12-22, and for a more idiosyncratic view of the meeting see Pascoe, J., 2006. *The hummingbird cabinet. A rare and curious history of romantic collectors*. Ithaca, New York: Cornell University Press, 140-168.

44. The Royal Society list also included Caroline Herschel (1750-1848), Mary Somerville (1780-1872), Elizabeth Garrett Anderson (1836-1917), Hertha Ayrton (1854-1923), Kathleen Lonsdale FRS (1903-1971), Elsie Widdowson FRS (1908-2000), Dorothy Hodgkin FRS (1910-1994), Rosalind Franklin (1920-1958) and Anne McLaren FRS (1927-2007). The Bank of England shortlist, from 989 nominations, also included Paul Dirac, Rosalind Franklin, William Herschel and Caroline Herschel, Dorothy Hodgkin, Ada Lovelace and Charles Babbage, Stephen Hawking, James Clerk Maxwell, Srinivasa Ramanujan, Ernest Rutherford, Frederick Sanger and Alan Turing. In July 2019 it was announced that Alan Turing had been selected to appear on the new polymer note. For examples of using Anning in teaching see Harnett, P. & Whitehouse, S., 2017. Investigating activities using sources at Key Stage 1. *In* Cooper, H. (ed.), *Teaching history creatively*. Second Edition. Abingdon: Routledge, 39; and Roden, J. & Archer, J., 2014. *Primary science for trainee teachers*. London: SAGE Publications, 99. See also Clary, R.M. & Wandersee, J.H., 2006. Mary Anning: she's more than "seller of sea shells at the seashore". *The American Biology Teacher*, v.68, 153-157.

45. Mary Anning Rocks. *Geoscientist*, v.30, May 2020, 26.

46. Purcell, R.W. & Gould, S.J., 1992. *Finders, keepers. Eight collectors*. London: Hutchinson Radius, 100. The assessment that more has been written about Anning than any other British geologist is by Oldroyd, D., 2013. A preliminary analysis of biographies and autobiographies of geologists. *Earth Sciences History*, v.32, vii–xii.

47. The new Golden Age of marine reptile research is from Taylor, M.A., 1997. Before the dinosaur: the historical significance of the fossil marine reptiles. *In* Callaway, J.M. & Nicholls, E.L. (eds). *Ancient Marine Reptiles*. San Diego: Academic Press. For more on modern research on marine reptiles see also Elllis, R., 2003. *Sea Dragons. Predators of the Prehistoric Oceans*. Lawrence, Kansas: University of Kansas Press; Lomax, D.R., 2018. Fossils explained 72. Hidden sea dragons – discovering new species of Jurassic ichthyosaurs in museum collections. *Geology Today*, v.34, 236-240. An interesting comparison of Anning with other, more recent fossil collectors is in Noè, L., Gómez-Pérez, M. & Nicholls, R., 2019. Mary Anning, Alfred Nicholson Leeds and Steve Etches. Comparing the three most important UK 'amateur' fossil collectors and their collections. *Proceedings of the Geologists' Association*, v.130, 366-389. For more on women in science in Anning's time see Orr, M., 2015. Women peers in the scientific realm: Sarah Bowdich (Lee)'s expert collaborations with Georges Cuvier, 1825-33. *Notes and Records of the Royal Society of London*, v.69, 37-51; Patterson, E.C., 1983. *Mary*

Somerville and the cultivation of science 1815-1840. International Archives of the History of Ideas. The Hague: Martinus Nijhoff Publishers; and Burek, C.V. & Higgs, B. (eds), 2007. *The role of women in the history of geology.* Geological Society Special Publication 281, London: The Geological Society. The History of Geology Group of the Geological Society was founded in 1994.

48. 'This persevering and successful *collectress* of extraneous fossils' is from Anon., 1829. Organised fossil remains at Lyme, Dorsetshire. *The Crypt, or receptacle for things past, and West of England Magazine,* v.1 part 1, January to June 1829, 163-164.

MARY ANNING'S FOSSILS

1 For details of Anning specimens in the Natural History Museum, London, see Chapman, S. & Milner, A.C., 2010. Appendix. A catalogue of fossil specimens collected by Mary Anning (1799-1847) held in the collections at the Natural History Museum, London. *In* Lord, A.R. & Davis, P.G. (eds), 2010. *Fossils from the Lower Lias of the Dorset coast.* Palaeontological Association Field Guide to Fossils: Number 13. London: The Palaeontological Association, 401-407. The 1823 *Plesiosaurus dolichodeirus* is NHMUK 22656; the 1830 *Plesiosaurus macrocephalus* is NHMUK R1336; and the 1828 *Dimorphodon macronyx* is NHMUK R 1034.

2. Lang, W.D., 1945. Three letters by Mary Anning, Fossilist of Lyme. *Proceedings of the Dorset Natural History & Archaeological Society,* v.66, 169-173. The Anning ichthyosaur skull, museum registration no. NHMUK R1158, was identified in Delair, J.B., A history of the early discoveries of Liassic ichthyosaurs in Dorset and Somerset (1779-1835). *Proceedings of the Dorset Natural History & Archaeological Society,* v.90, 115-127. The missing skeleton may have been lost early on; Richard Lydekker in his 1889 *Catalogue of the fossil Reptilia and Amphibia in the British Museum (Natural History),* Part 2. London: British Museum (Natural History), 96, describes the skull and the attached part of the skeleton but not the remainder of the specimen. Nor, curiously, is it mentioned in William Bullock's 1816 *A companion to the London Museum, and Pantherion, containing a brief description of upwards of fifteen thousand natural and foreign curiosities, antiquities, and productions of the fine arts; now open for public inspection in The Egyptian Temple, Piccadilly, London.* 17th edition. London: Printed for the Proprietor.

3. Anning specimens in Oxford include the cephalopod *Phragmoteuthis huxleyi,* J.03578 and J.03532; *Phragmoteuthis montefiorei,* J.23739; a belemnite ink-sac from the Lias of Watchet in Somerset, J.03564, but although this may have been acquired from Anning, it may not have been collected by her; a coprolite presented by Anning, J.023781; a plesiosaur paddle, J.50146; and *Ichthyosaurus anningae,*

NOTES, SOURCES AND FURTHER READING

J.13587. I am grateful to Chris Duffin and Eliza Howlett for information about the Oxford collection.

4. John Philpot (1809-1878) was a London solicitor like his father, also John Philpot (1778-1850). The tail of *Squaloraja polyspondyla* is J.3097. Details of the important (type) specimens in the Philpot Collection are in Edmonds, J.M., 1978. The fossil collection of the Misses Philpot of Lyme Regis. *Proceedings of the Dorset Natural History & Archaeological Society*, v.98, 43-48.

5. Sedgwick, A., 1869. Prefatory Notice. *In* Seeley, H.G., *Index to the Fossil Remains of Aves, Ornithosauria and Reptilia, from the Secondary System of Strata arranged in the Woodwardian Museum of the University of Cambridge*. Cambridge: Deighton, Bell, and Co., iv.

6. For more on the Cambridge Anning specimens see Price, D., 1986. Mary Anning specimens in the Sedgwick Museum, Cambridge. *The Geological Curator*, v.4, 319-324. The ichthyosaur identified by a sketch in Anning's letter of 4 May 1843 has the museum's registration number J.35189. The other probable Anning ichthyosaur specimens are a skull with six vertebrae, J.59645, mentioned in a letter of 11 February 1831; a skull, J.68446, mentioned on 4 and 20 May 1843; a skeleton in a letter of 29 June 1835, J.35187; another mentioned on 11 February 1831, J.59642; and a skeleton in a letter of 20 May 1832 now identified as *Ichthyosaurus breviceps*, 50187. The 'beautiful Pentacrinite' mentioned by Sedgwick in 1869 could be one of several specimens in the museum.

7. See Massare, J.A. & Lomax, D.R., 2014. An *Ichthyosaurus breviceps* collected by Mary Anning: new information on the species. *Geological Magazine*, v.151, 21-28. This specimen has the number CAMSMX. 50187.

8. Charles Moore (1815-1881); Robert Milson Appleby (1922-2004); William Henry Eastwick (1780-1854). The large *Temnodontosaurus* skull has the registration number M3577a. For more on Moore and his collection see Copp, C.J.T., 1975. The Charles Moore Geological Collection. *Newsletter of the Geological Curators' Group*, v.1, 109-116; Copp, C.J.T., Taylor, M.A. & Thackray, J.C., 1997. Charles Moore (1814-1881). *Proceedings of the Somerset Archaeology and Natural History Society*, v.147, 1-36; Torrens, H.S., 2010. Uncurated Curators, No.3. Ronald Frederick Pickford (1920-2010): Bath curator, a tribute. *The Geological Curator*, v.9, 243-254.

9. John Templeman (c.1757-1848) lived at 33 Pulteney Street in Bath. See Torrens, H.S., 1975. The Bath geological collections. Introduction [and] Alphabetical listing of the major geological donations to the Bath Museum. *Newsletter of the Geological Curators Group*, v.1, 88-108. Templeman's ichthyosaur may be *Ichthyosaurus breviceps*, rather than *I. communis*, and has the registration number M3570.

10. The ammonite has the registration number TTNCM: 8517. I am grateful to Amal Khreisheh of Somerset Museums Service for information on this specimen.

11. Knell, S. J., 1986. Some little known collections in the South East. CING 20. Saffron Walden Museum. *The Geological Curator*, v.4, 351-352. Wyatt George Gibson (1790-1862).

12. John Naish Sanders (c. 1777-1870). The specimen now has the registration number Cc 940. For more on this specimen and the other large ichthyosaur skulls see Torrens, H.S., 2008. A saw for a jaw. *Geoscientist*, v.18,18-21.

13. Torrens, H.S., 2008. A saw for a jaw. *Geoscientist*, v.18,18-21.

14. William Henry Fitton (1780-1861). This event was recorded in a journal kept by the Geological Society's clerk and which is in the Society's archives, LDGSL/282. I am grateful to the Society's Archivist, Caroline Lam, for bringing this to my attention.

15. James Luke (1799-1881). Owen, R., 1854. *Descriptive Catalogue of the Fossil Organic Remains of Reptilia and Pisces contained in the Museum of the Royal College of Surgeons of England*. London. That four specimens survived is stated by Delair, J.B., 1969. A history of the early discoveries of Liassic ichthyosaurs in Dorset and Somerset (1779-1835). *Proceedings of the Dorset Natural History and Archaeological Society*, v.90, 115-127. I am grateful to Sarah Pearson of the Royal College of Surgeons for details of the four surviving fossil reptiles, one of which was a plesiosaur vertebra from Oxfordshire presented by William Buckland while the other three are from John Hunter's collection.

16. I am grateful to Simon Harris at the British Geological Survey for this information about the Anning specimens at BGS. Murchison succeeded De la Beche as Director in 1855 and remained in post until his death in 1871.

17. Edmund Thomas Higgins (1817-1891). Higgins' donation of the ichthyosaur is noted in the *Annual Report of the Council of the Yorkshire Philosophical Society for* 1848, 10, 22. I am grateful to Mike Simms of the National Museums of Northern Ireland for details of its acquisition by the Ulster Museum. Its registration number is BELUM K1296.

18. For more on the Anning specimens in Paris see Vincent, P. & Taquet, P., 2010. A plesiosaur specimen from the Lias of Lyme Regis: the second ever discovered plesiosaur by Mary Anning. *Geodiversitas*, v.32, 377-390; and Vincent, P., Taquet, P., Fischer, V., Bardet, N., Falconnet, J. & Godefroit, P., 2014. Mary Anning's legacy to French vertebrate palaeontology, *Geological Magazine*, v.151, 7-20. The Anning-Birch ichthyosaur specimens include a skull of *I*. cf. *communis* MNHN AC 9862 and a partial skeleton of *Leptonectes tenuirostris* MNHN AC 9937. The Waring-Anning plesiosaur, *Plesiosaurus dolichodeirus,* is MNHN AC 8592.

19. Prince Albert of Sachsen-Coburg and Gotha (1819-1861) was elected a Fellow of the Geological Society at a meeting on 30 May 1849. For more on this collection, see Mönnig, E., 2018. COBURG: Naturkunde-Museum Coburg – Paleontological Collections. *In* Beck, L.A. & Joger, U. (eds), *Paleontological*

NOTES, SOURCES AND FURTHER READING

Collections of Germany, Austria and Switzerland. The history of life of fossil organisms at museums and universities. Natural History Collections. Cham: Springer, 137-145; and Mönnig, E., 2007. Prinz Albert von Sachsen-Coburg und Gotha und die Naturkunde. *In* Bosbach, F. & Davis J. (eds). *Windsor-Coburg. Geteilter Nachlass – Gemeinsames Erbe. Eine Dynastie und ihre Sammlungen,* 6 Abb. Munchen: KG Saur, 115-132.

20. Thomas Bellerby Wilson (1807-1865); Edward Wilson (1808-1888). See Spamer, E.E., Daeschler, E. & Vostreys-Shapiro, L.G., 1995. A study of fossil vertebrate types in the Academy of Natural Sciences of Philadelphia. Taxonomic, systematic, and historical perspectives. *The Academy of Natural Sciences of Philadelphia Special Publication* 16, 12. The donation is recorded as 'Skeleton of Ichthyosaurus 6 feet 7 inches from the tip of the rostrum to the end of the tail. From Lyme Regis, England. Skeleton of Plesiosaurus dolichodeirus? measuring 6 feet 4 inches from tip of snout to end of tail. From Lyme Regis.' in *Proceedings of the Academy of Natural Science of Philadelphia,* v.3, 1846 & 1847, 195-196.

Acknowledgements

Many people have been generous in their assistance while I have been working on this book, especially during lockdown and furlough. I am grateful for the help I have received from so many friends and colleagues and from fellow curators, librarians, archivists and geologists who have uncomplainingly responded to yet another email enquiry or visit from me and for allowing me to quote from archives in their care. Thank you to Indy Bhullar, LSE Library; Cynthia Burek, University of Chester; Richard Edmonds, Lyme Regis; Simon Harris, Mike Howe, Andrew Morrison and Paul Shepherd, British Geological Survey; John Henry; Cindy Howells and John Cope, National Museum of Wales; Eliza Howlett, Kate Diston, and Wendy Shepherd, Oxford University Museum of Natural History; Sarah Kenyon, Saffron Walden Museum; Amal Khreisheh, Somerset Museums Service; Sarah King, Yorkshire Museum; Caroline Lam, Wendy Cawthorne and Michael McKimm, Geological Society Library; Cherry Lewis; Susan Newell; Sarah Pearson, Royal College of Surgeons, London; Holly Peel and Amelia Walker, Wellcome Collection; Hellen Pethers, Emma Bernard, Chris Duffin and Stephen Atkinson, Natural History Museum; Christopher Powell; Liz Selby, Dorset County Museum; Mike Simms, National Museums of Northern Ireland; Martin Simpson; David Tomalin; David Tucker, Lyme Regis Museum; Hannah Clarke, University of Aberdeen; Alexander Turnbull Library, Wellington, New Zealand; and Matt Williams, Bath Royal Literary and Scientific Institution.

Particular thanks are due to Ildi Clarke for her index, Peter Lightfoot for the map on page 6, John Marriage for photographing material in Lyme Regis Museum, and to Charlotte Topsfield and David Burnett for their insightful comments on and improvements to the text, as well as to Helen Field and Alastair Macrae for taking the time to read through early drafts. I very much appreciate their help. Special thanks, too, to Wendy Cawthorne and Caroline Lam of the Geological Society for chasing up material in the Society's collections, finding curious snippets, and pursuing Mary Anning in London.

Most of all I am indebted to Hugh Torrens, Mike Taylor and Richard Bull who have been a mine of information on Mary Anning and the history of Lyme Regis, and more than willing to share their extensive knowledge, and to David Burnett of The Dovecote Press for his invitation to write this book.

TOM SHARPE *September 2020*

The Illustrations

I would like to express my gratitude to the following for their kind permission to use illustrations from their collections or for which they hold the copyright: University of Aberdeen, 34; Amgueddfa Cymru - National Museum Wales, 26, 48, 58; Richard Bizley, 59; Richard Bull, 7, 79; Cindy Howells/Bath Royal Literary and Scientific Institution, 76; The Geological Society, 25, 45, 66; Lyme Regis Museum, 5, 6, 38, 41, 50, 54, 81; Natural History Museum, London, 23, 24, 27, 35, 37, 43, 46, 49, 52, 61, 62, 72, 74; Oxford University Museum of Natural History, 69, 73, 75; Private Collection, 30, 40, 71, 82; Wellcome Collection. CC BY, frontispiece, 28, 32, 33, 36, 64, 70. All the remaining illustrations come from the collections of the author and the Dovecote Press.

Index

Academy of Natural Sciences, Philadelphia, USA, 92, 121, 172
Agassiz, Louis, 41, 101, 104, 105, 106, 121, 126-8, 133, 134, 141, 144
Albert, Prince, of Sachsen-Coburg and Gotha, 171
Allan, Thomas, 70, 71
ammonites, 11, 17, 20, 29, 31, 33, 34, 35, 43, 46, 50-1, 52, 56, 61, 73, 102, 108, 111, 132, 137, 138, 149, 155, 159, 164, 167
   Ammonites Bucklandi, 137
   ink-sacs in, 137
   Ludwigia murchisonae, 94
   Trigonellites/Aptychus, 137
Anning, Albert, 150, 151
Anning, Amelia, 145, 154, 155
Anning, Anne née Flood, 24
Anning, Annette, 154
Anning, Charles Churchill, 34, 35, 37
Anning, Joseph, 24, 25, 28, 32, 34-5, 37, 40, 43, 56, 58-9, 60, 67, 77, 107, 125, 129, 140, 144, 145, 153, 154, 155
Anning, Mary
   Acrodus anningiae, 128
   Anninga megalops, 157
   Anningasaura, 158
   Anningella, 157
   Belenostomus anningiae, 128
   celebrity of, 26, 46, 107, 115-16, 132-3, 146, 148, 157, 163
   Cytherelloidea anningi, 157
   Duria antiquior plate, 102-3, 128, 142
   education of, 25-6
   family of, 23, 24-8
   'fossil-extractor' of, 153
   'Fossil Shop', 37, 56, 61, 78, 82ff., 103, 108, 126, 132, 141, 143, 149
   Ichthyosaurus anningae, 165
   Kenyon's poem, 132-3
   legends/myths about, 156-61
   lightning strike story, 9, 26-7, 35, 61, 72, 114, 132, 145, 159
   museum collections of, 164-72
   portraits of, 154-6
   scientific recognition of, 7-8, 42, 58, 68, 69-71, 85, 92, 106, 116, 132-3, 141-2, 143, 145-6, 157-8, 159-60, 161-3, 164
   Tricycloseris anningi, 157
Anning, Mary (Molly) née Moores, 24, 25, 26, 32, 34, 35, 37, 55, 59-60, 78, 107, 118, 128, 137-8
Anning, Richard, 15, 24, 25, 26, 27, 28, 29-30, 31-2, 34, 36, 46, 137, 145, 159
Anning, William, 24
Aplysia punctate, 86
Appleby, Dr Bob, 167
Asteroceras obtusum, 51
Austen, Jane, 8, 15, 120
Axminster, 22, 48, 56, 98, 134

Bakewell, Robert, 37
Bank of England, 8
Bard, Wilkie, 161
Bath, 14, 15, 21, 22, 24, 36, 57, 83
Bath Royal Literary and Scientific Institution, 167
belemnites, 17, 20, 30, 35, 43, 46, 56, 61, 73, 84-6, 124, 150
   sepia in, 85-6, 137
   squid, relation of, 84-5
   See also cephalopods
belemnon, 84
Belemnosepia, 86
Bell, Alexander Melville, 161
Bell, Frances Augusta (Fanny), 33, 71-6, 95, 114, 115, 117
Benett, Etheldred, 43, 121-2, 123
Bennett, Captain Charles, 77
bezoar stones, 86-7, 92
Birch, Col. Thomas James, 53, 54-5, 57, 58, 65, 68, 83, 118, 120, 128, 151, 170, 171
bivalves, 35, 164
Blandford Forum, 23, 24
Blumenbach, Johann Friedrich, 83
Botfield, Beriah, 105, 106
brachiopods, 108
Braikenridge, George Weare, 56
Bright, Richard, 57
Brighton, 13, 14, 15
Bristol, 13, 22, 24, 48, 49, 56, 57, 66, 77, 101, 168
Bristol City Museum, 168, 170
Bristol Institution for the Advancement of Science and Art, 60, 77, 80, 91, 99, 100, 101, 105, 127, 166, 168, 169
Bristol Library Society, 58
Bristol Philosophical Society, 66, 77
British Association for the Advancement of Science, 88, 128-9, 134, 150
British Geological Survey, 45, 94, 134, 145, 149, 169, 170, 171
   Anning specimens in, 170
British Museum, 8, 37, 44, 59-60, 64, 65, 73-4, 85, 90, 91, 92, 95, 96, 113, 114, 121, 131, 136, 141, 151, 157, 158, 164, 168, 169
   Anning's ichthyosaur in, 73-4, 136, 165
   Anning's plesiosaurus in, 92, 98

INDEX

Broderip, William John, 105, 116
Broom, Robert, 157
Brown, Henry Rowland, 148, 158
Buckland, Rev. Charles, 46
Buckland, Elizabeth, 46
Buckland, Frank, 46, 88, 158
Buckland, Mary née Morland, 42, 48, 50, 85, 118, 121, 124, 128, 135
Buckland, William, 7, 11, 19, 22, 39, 41, 44, 45-8, 49ff., 57, 59, 61, 65-6, 67-8, 69, 74, 82, 84-92, 98, 99, 100, 102-4, 105ff., 112, 113, 116, 117, 118, 121, 126, 128-9, 134, 135, 143, 144, 146, 147, 151, 152, 156, 158, 165, 170, 171
Bull, Richard, 153
Buller, William, Bishop of Exeter, 107
Bullock, William, 37-8, 44, 54, 55, 56, 73, 136, 165
Bunbury, Sir Henry, 65
Burek, Cynthia, 121

Cardiff University, 167
Carpenter, Dr Thomas, 31, 39, 43, 61
Carus, Carl Gustav, 140-1
Catcott, Rev. Alexander, 58
Cautley, Proby Thomas, 150
cephalopods, 84, 85
  sepia (ink) in, 85
Chantrey, Sir Francis, 85
Chard, Henry, 13-14
Charlesworth, Edward, 133, 134
Charmouth, 8, 11, 19, 20, 22, 25, 28-9, 31, 37, 43, 51, 56, 77, 84, 109, 111, 126, 149-50, 154, 157
  fossil collectors/shops in, 148, 149-50
  submarine forest discovery, 77
Chevalier, Tracy, 120, 160
Clark, Samuel, 150
Clark, William, 150
Clarke, Thomas, 58
Clift, William, 134
Cobbold, Elizabeth, 121
Cole, William Willoughby, 3rd

Earl Enniskillen, 104-5, 128, 147, 150-1, 155, 164, 165
  Coleia fossil, 105
  Plesiosaurus macrocephalus, 104, 105, 128
conchology, 79
Congreve, Ann, 51, 121, 123
Congreve, Mary, 51
Conybeare, William Daniel, 22, 45, 47, 48-9, 56, 57, 58, 60, 61, 64, 65-9, 71, 77, 80, 98, 110, 112, 113, 116, 121, 122-3, 126, 134, 135, 139, 146, 147, 151, 152, 157, 168, 170, 171
Cooper, Sir Astley, 106
coprolites, 7, 86-8, 98, 100, 104, 107, 113, 144, 165
  See also bezoar stones
Cowstones, 97
Cox, Leslie Reginald, 157
crinoids, 17, 51, 52, 82-3, 90-1, 105, 107, 130, 164, 167, 168
  Pentacrinites, 83, 90-1, 92, 130
  Pentacrinus briareus, 167
Crook, Rev. John, 25
Crookshanks, John, 30
Crowther, Peter, 168
Crystal Palace, 160
Culverhole Point, 106-7
Cumberland, George, 55, 57, 60, 68, 118, 123, 168
Cumming, Lady Eliza Maria Gordon, 121
Cunnington, William, 28, 29
Curtis, Sidney, 149
Cuvier, George, 40, 41, 65, 68-9, 92, 96, 100, 141, 171

Dapedium, 30, 36, 83, 167, 170
  Dapedium politum, 170
Darwin, Charles, 46, 55, 56, 126, 141, 146, 157, 163
Davis, Mary Anne, 72
Davy, Sir Humphrey, 94
Dawson, William, 135
Day, Samuel Skurray, 57
De la Beche, Sir Henry, 22, 37, 39, 42-5, 47, 49, 50, 51, 53, 55, 56-8, 59ff., 64ff., 71, 77, 79, 86, 94, 96, 97, 98, 102-3,

105, 106, 110, 112, 113, 116, 117, 118, 119-20, 121, 128, 133, 134, 140, 141-2, 145, 146, 147, 149, 151, 156, 157, 168-9, 170
  Duria antiquior lithograph, 102-3, 128, 142
  Hybodus Delabecheii, 134, 170
De Luc, Jean André, 31-2
Dibdin, Thomas Frognall, 96
Dickens, Charles, 158
dinosaurs, naming/classification, 47, 68, 160
Dissenters, 25, 28, 125
Donne, Benjamin John Merefield, 155
Dorchester, 143
Dorset County Museum, 143
Dorset Natural History and Archaeological Society, 8
Draper, Jo, 152
driftwood, 19, 91
Duckworth, Admiral Sir John Thomas, 107
Duckworth, Lady Susannah-Catherine, 107
Dunster, Daniel, 135
Durweston, 23

Eastwick, William Henry, 167
echinoderms, 17, 82
Edward I, 12
Edward VII, 37
Edwards, James, 109
Egerton, Philip de Malpas Grey, 104, 116, 129, 130, 147, 151, 165
Egyptian Hall, London, 37, 54
evolution, 56, 126
Exeter, 107
extinction, 40-1

Fagan, Rev. Henry Stuart, 158-9
Fane, Henry MP, 13
Fane, John, 9th Earl of Westmorland, 13
Featherstonhaugh, George William, 82, 91, 171
Feltham, John, 14, 15
Fielding, Henry, 109
Fisher, Rev. Osmond, 138, 143
Fitton, William Henry, 169

237

Forde, Harriet Anne, 159
fossil collecting/selling, 14-15, 19-20, 28-30, 42, 43-4, 50, 52, 54-5, 61-4, 79, 148-50
fossil fish, 126-8, 133-4, 136, 151, 170
  Acrodus anningiae, 128
  Belenostomus anningiae, 128
  Eugnathus philpotiae, 127
  See also *Dapedium*; *Hybodus*
fossils, 10-11, 17-20, 35-6
  Biblical explanation of, 20, 41, 47, 126
  first illustrations of, 36
  formation of, 17-18, 20-1, 126
  Greensand fossils, 96-7
  monetary values of, 10-11
  names given to, 11
  wood/lignite, 90-1
  See also driftwood; geology
Fowles, John, 13, 152, 161
Frederick Augustus II, 140-1

Galpin, Thomas Carter, 156
gastropods, 17, 35
Geological Society, 22, 43, 44, 45, 46, 55, 57, 66, 68, 77, 82, 84, 86, 87, 89, 94, 95-6, 98, 101, 105, 109, 110, 116, 122, 127, 129, 136-7, 143, 145, 147, 154, 155, 163, 168-9, 172
  Anning specimens in, 168-9
  women's involvement in, 122, 145, 160
Geological Survey see British Geological Survey
Geologists' Association, 122, 150
geology, 9, 17, 20, 21, 26, 40, 41, 94, 160, 163
  Belemnite Marl rocks, 84
  Blue Lias Formation, 16-17, 35-6
  Cambrian Period, 55
  Charmouth Mudstone Formation, 11, 17
  Cretaceous Period, 17
  dating of rocks, 21
  Devonian System, 56, 94
  female contribution to, 121-3, 160, 163
  Jurassic Period, 16, 21
  Permian System, 94

Silurian System, 94
George III, 14
George IV, 14
Gibson, Jebez, 168
Gibson, Wyatt George, 168
Gifford, Harry, 161
Gleed, Rev. John, 77, 125
Godwin, Phil, 12
Goodhue, Tom, 25-6, 132
Goodrich, Samuel Griswold, 142
Gould, Stephen Jay, 112, 162
Graham, Maria, 122
Grant, Rev. Johnson, 76
Grant, Dr Robert, 96
Gray, William, 154, 155
Grenville, Richard 1st Duke of Buckingham, 65, 66, 69, 71, 80, 91, 100
Grenville, William, 1st Baron Grenville, 68

Hack, Maria, 157
Hallett, Salina, 50
Harcourt, William Vernon, 129
Haskings, Elizabeth, 26
Haskings, John, 26
Hawker, Rev. Peter, 36, 39, 44
Hawkins, Thomas, 108-14, 116, 125, 128, 136, 147, 157, 164, 165, 167
  ichthyosaur excavation, 109-10, 131
Henley, Henry Hoste, 37-8, 62, 73
Higgins, Edmund Thomas, 133, 170, 171
Hodges, Rev. Frederic Parry, 125-6, 144, 146
Hodges, Rev. Thomas, 126, 129, 143
Holland, George, 43
Hollis, Thomas, 13, 15
Home, Sir Everard, 38, 39-40, 42, 44, 52, 55, 57, 66, 67, 68, 74, 85, 157, 170
Howman, Rev. George Ernest, 90
Hullmandel, Charles, 102
Hunter, Isaac, 150
Hunter, John, 149-50, 170
Hutton, James, 21, 41
Hutton, William, 110
Hybodus, 133-4, 136

Hybodus delabechei, 134, 170

Ichthyosaurus, 10, 11, 17, 30, 34-5, 36-40, 44, 51, 52, 55, 57, 58-9, 62, 65, 67, 68, 69, 71, 72-3, 79-80, 85, 86, 89, 99, 101, 102, 105-6, 109-10, 125, 130-1, 136, 138, 141, 146, 151, 157, 164, 165, 166-7, 168-9, 171, 172
  Anning's first discovery of, 34-5, 36, 43, 44, 73, 74, 131, 159, 165
  bezoar stones/coprolites, 86-7, 88
  as fossil 'crocodile', 30, 34, 35, 36, 37-9, 40, 44, 56, 57
  Ichthyosaurus anningae, 165
  Ichthyosaurus breviceps, 166-7
  Ichthyosaurus communis, 11, 45, 59, 79-80, 165, 167, 171
  Ichthyosaurus platyodon, 45, 110, 113, 131, 138, 139, 148
  Proteosaurus, 39-40, 57, 170
  scientific naming of, 34, 36, 39-40, 44, 45, 57, 110, 157, 165
  species naming/recognition of, 45, 57, 112
  Temnodontosaurus platyodon, 110, 165, 167, 168-9

Johnson, James, 28, 29, 30, 37, 39, 44, 53, 56, 61, 168-9
Jurassic Coast World Heritage Site, 17

Kennaway, Robert, 51
Kennaway, Sarah, 42, 51-3, 56, 126, 128, 144
Kerr, Lady Isobel, 90
Kerr, John, 7th Marquis of Lothian, 90
Kidd, Dr John, 46, 48, 49
Kirkdale Cavern, Yorkshire, 41, 87, 98
Konig, Charles, 37, 44, 59, 60, 74, 91, 92, 114, 141

Lamb, William, Lord Melbourne, 129

Lang, William Dickson, 8, 108-9, 159, 160, 165
Lee, Francis, 120
Leichhardt, Ludwig, 131-2
Leptonectes tenuirostris, 60
lignite, 90-1
Lloyd, William, 29
Lock, William, 29, 36
Lomax, Dean, 166
Lonsdale, William, 95
Luke, James, 170
Lyell, Charles, 31, 41, 46-7, 48, 67, 68, 69, 70, 90, 99, 100, 102, 106, 129, 152
Lyme Regis, 7, 8, 11, 12-15, 19-20, 21-2, 23ff., 61, 72, 78, 124, 135, 140, 149, 152-3, 156, 162, 163
  Assembly Rooms, 15, 45, 76
  Church Cliffs, 12, 28, 30-1, 34, 62, 63, 89, 109, 144, 163
  the Cobb, 12, 13-14, 17, 27, 31, 63, 67, 72, 76
  commemoration of Mary Anning in, 162
  Dinosaurland (Independent Chapel), 25
  fire of 1844, 140
  geology of, 16-17
  great storm of 1824, 76-7
  history of, 12-15, 21-2, 61
  landslip (1939), 135
  Moore's Fossil Depot, 149
  quarrying in, 16, 28, 63, 109
  Rack Field, 26, 27
  sea-bathing in, 13, 14, 31, 71
  St Michaels, 30, 125, 144, 146, 155, 162
Lyme Regis Fossil Festival, 162
Lyme Regis Museum, 88, 90, 149, 152-3, 156

Mackintosh, Charles Rennie, 149
mammoths, 56
Mantell, Gideon, 41, 46, 48, 54, 67, 68, 69, 70, 78, 83, 100, 102, 106, 108, 109, 112, 113, 136, 157
Marder, Henry, 148
Marder, James, 148, 155
Martin, John, 112, 113
Massare, Judy, 166
Maton, Dr William George, 29

Megalosaurus, 47, 68
Melton Mowbray, 16
Miller, J. S., 51, 77, 99, 151
Montgomery, James, 120-1
Moore, Charles, 167
Moore, Sir John, 93
Moore, Sarah, 149
Moore, William, 148, 149
Morgan, William, 57
Murchison, Charlotte née Hugonin, 42, 93-6, 97, 98, 99, 103, 104, 110, 111, 113, 117, 120, 121, 124, 144, 170
Murchison, Sir Roderick, 56, 93-6, 97, 110-11, 121, 129, 136, 144, 147, 152, 170
Murray, John, 62, 150
Muséum National d'Histoire Naturelle (Paris), 40, 55, 69, 164, 171
Museum of Economic Geology, 149
Museum of Practical Geology, 170

National Archives, 10
National Museum of Wales, 167
National Museums of Northern Ireland, Belfast, 171
National Science and Media Museum (Bradford), 156
Natural History Museum, 35, 60, 71, 105, 106, 110, 113, 136, 139, 151, 154, 155, 164, 165
  Anning's specimens in, 165
Naturkunde-Museum, Coburg, 171
Nautilus, 56
New York Lyceum of Natural History, 171-2
Nightingale, Florence, 8
NonConformists, 25

Ornithorhynchus, 39
Owen, Richard, 90, 98, 116, 121, 134, 141, 144, 147, 151, 157, 160, 170
Oxford Geology Club, 49
Oxford University Museum of Natural History, 101, 106, 152, 164, 165-6
  Anning's specimens in, 165-6

Philpot sisters contributions in, 165-6

Palaeontographical Society, 150
Palaeontological Association, 162
  Mary Anning Award, 162
palaeontology, 16, 20, 22, 58, 87, 122, 157, 160, 162-3
  recognition of Mary Anning in, 162-3
  See also fossils; geology
Parkinson, James, 82
Parsons, Mary, 3rd Countess of Rosse, 121
Paterson, Lady Margaret, 107
Paterson, Major-General William, 107
Pearce, Robert, 131
Pentacrinites, 83, 90-1, 166, 167, 168
  See also crinoids
Perry, Sarah, 160
Phillips, John, 150
Phillips, William, 49, 139
Philpot, Elizabeth, 42, 48, 49, 50, 51, 52, 61, 71, 81, 85, 86, 96, 107, 121, 123, 124, 126, 127, 128, 134, 144, 147, 165-6
  Eugnathus philpotiae, 127
Philpot, Elizabeth Mary, 166
Philpot, John, 49, 51, 81, 166
Philpot, Margaret, 42, 49
Philpot, Mary, 42, 49
Philpot, Thomas, Embray Davenport, 152
Pidgeon, Edward, 116
Pinney, Anna Maria, 33, 34, 63, 114-18, 125, 129
Pinney, William, 114, 129
Plesiosaurus, 17, 30, 36, 56, 58, 64-9, 71, 77, 80, 87, 89, 91-2, 96, 102, 141, 146, 151, 162, 165, 171, 172
  Anningasaura, 158
  discoveries and naming of, 58, 64-9, 91-2, 97-8, 99, 103-4
  link with crocodiles, 58
  Plesiosaurus dolichodeirus, 11, 91, 104, 108, 165
  Plesiosaurus macrocephalus, 104, 105, 165
Postgate, Oliver, 90

Preston, J. W., 159
Prévost, Constant, 69-70, 90, 171
Price, David, 166
Prout, William, 87
Pterodactylus macronyx, 89
  Anning's discovery of, 88-9, 90, 96, 97, 98, 165
  as *Dimorphodon macronyx*, 90, 165
Pterosaurs, 17, 88-90, 92, 98, 100, 102, 157
  as flying dragons, 89, 90
Pugin, Augustus W. N., 146
pyrite, 131

Queen Victoria, 135, 152, 171

Rackett, Rev. Thomas, 139
Rattenbury, John, 23
Rawdon-Hastings, Barbara, Marchioness of Hastings, 121
Rawlins, Rev. Francis John, 126
Rawlins, Rev. Henry Williams, 126
Riley, Dr Henry, 101, 116
Roberts, George, 8, 14, 16, 26, 28, 29, 30, 31, 33, 34, 40, 59, 61, 62, 64, 78, 115, 129, 131, 135, 145, 147, 149, 150, 152, 154, 155, 156
Rowe, William, 89
Royal Academy, 154
Royal College of Surgeons, 38, 44, 52, 55, 57, 61, 96, 134, 141, 170
  Anning specimens in, 170
Royal Geographical Society, 94
Royal Society, 8, 39, 66, 67, 93, 94, 162
Russell, Richard, 13

Saffron Walden Museum, 168
Sandby, Paul, 93
Sanders, John Naish, 101, 168
Saull, William Devonshire, 106
Scharf, George, 102, 112
Sedgwick, Rev. Adam, 55-6, 81, 94, 95, 100, 104, 105, 126, 129-31, 138-9, 144, 146, 147, 152, 157, 166
Sedgwick Museum, University of Cambridge, 131, 139, 152,
164, 166-7
  Anning's specimens in, 166-7
Seward, Anna, 120
sharks *see Hybodus*
Shaw, Rev. Stebbing, 29
Sherborn, Charles Davies, 151
Sidbury, 24, 25
Silvester, Lady Harriet, 70-1
Silvester, Sir John, 70
Skinner, Charlotte Jane, 119
Smith, William, 21, 22, 82
smuggling, 14, 23, 117
Solly, Dorothea, 139, 140
Solly, Samuel, 139
Somerset Archaeological and Natural History Society, 167
Somerset County Museum, Taunton, 167
Somerville, Mary, 93
Sopwith, Thomas, 156
South, John, 29-30
Sowerby, George Brettingham, 79, 80-1, 82, 96
Sowerby, James, 11, 51, 52, 54, 79, 94, 97
Sowerby, James De Carle, 51, 94, 97
Squaloraja polyspondyla, 101, 105
  Anning's discovery of, 99-101, 127, 166, 168
starfish, 82, 105, 136, 165
  Ophiura egertoni, 105
Stock, Charlotte, 37, 42, 116-17
Stock, Dr John Edmonds, 37
Strickland, Hugh Edwin, 136-7
Studland Bay, 17
Stukeley, William, 58
Sullivan, Terry, 161

Talbot, William Henry Fox, 156
Taylor, Michael A., 9, 162
Taylor, Samuel Jr., 169
Templeman, John, 83, 167
Templeton, John Samuelson, 112
Thomas, Joan, 120, 160
Thorne, Isabel, 159
Tomes, Robert F., 157
Torrens, Hugh S., 9, 29, 70, 119, 159, 167, 168

Treves, Sir Frederick, 149
Tricycloseris anningi, 157
Tylee, Sarah, 121

Unity, 67, 77, 95, 131
University of Cambridge, 129, 131, 166
  See also Sedgwick Museum
University of London, 96
Ure, Andrew, 73-4
von Buch, Leopold, 142

Wailes, William, 146
Wanklyn, Cyril, 14, 152, 155
Warburton, Henry, 122
Waring, Eleanor Emma, 142-3
Waring, Capt. Henry, 69, 142, 171
Watchet, 106, 165
Wellesley, Sir Arthur, 1st Duke of Wellington, 14, 93
Weymouth, 13, 14
Whately, Richard, Archbishop of Dublin, 48
Wheaton, Rev. James, 25, 125
Whitby, 36, 58, 149
  Brown Marshall fossil shop, 149
Whitty, Rev. John, 25
Whyte, Letitia, 119
Wilkinson, Charles Hunnings, 44, 57
Williams, Rev. David, 100, 108
Wilson, Edward, 172
Wilson, Harriette, 14
Wilson, Bishop Thomas, 142
Wilson, Thomas Bellerby, 172
Winick, Stephen, 161
Wishcombe, Jonas, 111, 148
Wollaston, William, 87
Woods, Henry, 134
Woodward, Bernard Barham, 158
Woodward, Dr John, 129
Wordsworth, Dorothy, 23
Wordsworth, William, 23-4
Wyndham, Major-General Henry, 119
Wyse, Harriet, 155

Yorkshire Museum, 134, 171
Yorkshire Philosophical Society, 149, 171